Lecture Notes in Computer Science 7266

Alexey Vinel Rashid Mehmood
Marion Berbineau Cristina Rico Garcia
Chung-Ming Huang Naveen Chilamkurti (Eds.)

Communication Technologies for Vehicles

4th International Workshop, Nets4Cars/Nets4Trains 2012
Vilnius, Lithuania, April 25-27, 2012
Proceedings

Springer

Volume Editors

Alexey Vinel
Tampere University of Technology, 33101 Tampere, Finland
E-mail: alexey.vinel@tut.fi

Rashid Mehmood
School of Engineering, 8 Whitethorn Place, Swansea SA2 8HR, UK
E-mail: r.mehmood@gmail.com

Marion Berbineau
IFSTTAR, LEOST, 59650, Villeneuve d'Ascq, France
E-mail: marion.berbineau@ifsttar.fr

Cristina Rico Garcia
German Aerospace Center (DLR), 82234 Oberpfaffenhofen-Wessling, Germany
E-mail: cristina.ricogarcia@dlr.de

Chung-Ming Huang
National Cheng Kung University, Tainan, Taiwan
E-mail: huangcm@locust.csie.ncku.edu.tw

Naveen Chilamkurti
La Trobe University, Bundoora, Melbourne, VIC 3086, Australia
E-mail: n.chilamkurti@latrobe.edu.au

ISSN 0302-9743 e-ISSN 1611-3349
ISBN 978-3-642-29666-6 e-ISBN 978-3-642-29667-3
DOI 10.1007/978-3-642-29667-3
Springer Heidelberg Dordrecht London New York

Library of Congress Control Number: 2012935551

CR Subject Classification (1998): C.2, C.3, K.6.5, K.8.1

LNCS Sublibrary: SL 5 – Computer Communication Networks and Telecommunications

Typesetting: Camera-ready by author, data conversion by Scientific Publishing Services, Chennai, India

Printed on acid-free paper

Springer is part of Springer Science+Business Media (www.springer.com)

Preface

The Communication Technologies for Vehicles workshop series provides an international forum on latest technologies and research in the field of intra- and inter-vehicle communications and is organized annually to present original research results in all areas related to communication protocols and standards, mobility and traffic models, experimental and field operational testing, and performance analysis.

Previous Nets4Cars/Nets4Trains workshops were held in Saint Petersburg, Russia (2009), Newcastle, UK (2010), and Oberpfaffenhofen, Germany (2011). These proceedings contain the papers presented at the 4th International Workshop on Communication Technologies for Vehicles, which took place in Vilnius, Lithuania, in April 2012.

The workshop was collocated with the Second Baltic Conference on Future Internet Communications BCFIC-2012 (technically sponsored by IEEE) and meetings of two COST actions: IC0905 TERRA: Techno-Economic Regulatory framework for Radio spectrum Access for CR/SDR and IC0906 WiNeMO: Wireless Networking for Moving Objects.

Our call for papers resulted in 21 submissions. Each of them was assigned to the Technical Program Committee members, and 13 submissions were accepted for publication (9 for the road track and 4 for the rail track). Each accepted paper got at least three independent reviews. In addition, two invited papers were accepted. The order of the papers in these proceedings was aligned with the workshop program.

We extend a sincere "thank you" to all the authors who submitted the results of their recent work, to all the members of our hard-working comprehensive Technical Program Committee, as well as the thoughtful external reviewers.

April 2012

Alexey Vinel
Rashid Mehmood
Marion Berbineau
Cristina Rico Garcia
Chung-Ming Huang
Naveen Chilamkurti

Organization

Workshop Organizers

General Co-chairs

Alexey Vinel	Tampere University of Technology, Finland
Rashid Mehmood	UK

TPC Co-chairs (Nets4Trains)

Marion Berbineau	IFSTTAR, France
Cristina Rico Garcia	DLR, Germany

TPC Co-chairs (Nets4Cars)

Chung-Ming Huang	National Cheng Kung University, Taiwan
Naveen Chilamkurti	La Trobe University, Australia

Steering Committee

Xu Li	State University of New York at Buffalo, USA
Fei Liu	University of Twente, The Netherlands
Antonella Molinaro	University "Mediterranea" of Reggio Calabria, Italy
Joel Rodrigues	University of Beira Interior, Portugal
Juan de Dios Sanz Bobi	CITEF, Spain
Axel Sikora	University of Applied Sciences Offenburg, Germany
Thomas Strang	DLR, Germany
Tsutomu Tsuboi	Hitachi, Japan
Yan Zhang	Simula Research Laboratory, Norway

Technical Program Committee

Marina Aguado	University of the Basque Country, Spain
Onur Altintas	Toyota InfoTechnology Center, Japan
Atif Alvi	Lahore University of Management Sciences, Pakistan
Petros Belimpasakis	Bang & Olufsen, Germany
Marion Berbineau	IFSTTAR, LEOST, France
Mohamed Boucadair	France Telecom, France
Marcello Caleffi	University of Naples "Federico II", Italy
Marilia Curado	University of Coimbra, Portugal
Robil Daher	University of Rostock, Germany

Fouzia Elbahhar	INRETS, France
Thierry Ernst	Mines ParisTech, France
Fethi Filali	QU Wireless Innovations Center, Qatar
Francisco Garcia	Agilent Technologies / Strathclyde University, UK
Javier Goikoetxea	Construcciones y Auxiliar de Ferrocarriles, S.A., Spain
Javier Gozalvez	Universidad Miguel Hernandez de Elche, Spain
Christophe Gransart	INRETS, France
Oleg Gusikhin	Ford, USA
Andreas Lehner	DLR, Germany
Juliette Marais	IFSTTAR, LEOST, France
Francesca Martelli	IIT – CNR, Italy
Michael Meyer zu Hoerste	DLR, Germany
Markus Miche	SAP Research, Switzerland
David Mottier	Mitsubishi Electric R&D Centre Europe, France
John Murphy	University College Dublin, Ireland
Marc Necker	University of Stuttgart, Germany
Augusto Neto	Universidade Federal do Ceará, Brazil
Brian Park	University of Virginia, USA
Divitha Seetharamdoo	IFSTTAR, LEOST, France
Axel Sikora	University of Applied Sciences Offenburg, Germany
Markus Strassberger	BMW Group Research and Technology, Germany
Jouni Tervonen	University of Oulu, Finland
Bart van Arem	TU Delft, The Netherlands
Teresa Vazão	Inesc-ID/Instituto Superior Técnicom, Portugal
Martine Wahl	IFSTTAR, LEOST, France
Michelle Wetterwald	EURECOM, France
Nawaporn Wisitpongphan	KMUTNB, Thailand
Yunpeng Zang	RWTH Aachen, Germany
Yang Zhang	Palo Alto Networks, USA

Additional Reviewers

Yevgeni Koucheryavy	Tampere University of Technology, Finland
Jonathan Letessier	IPSIS, France
Jonathan Petit	University of Twente, The Netherlands
Olga Galinina	Tampere University of Technology, Finland
Boris Bellalta	UPF, Spain
Jacek Rak	Gdansk University of Technology, Poland
Timo Lehikoinen	VTT, Finland
Markus Pelz	DLR, Germany

Alexander Pyattaev	Tampere University of Technology, Finland
Jong-Hyouk Lee	INRIA, France
Wouter Klein Wolterink	University of Twente, The Netherlands
Jakub Jakubiak	Tampere University of Technology, Finland

Hosting Institution

Vilnius Gediminas Technical University, Lithuania

Table of Contents

Road Track

Rail Track

Cellular Networking Technologies in ITS Solutions: Opportunities and Challenges

Andreas Fasbender[1], Martin Gerdes[1], and Sascha Smets[2]

[1] Development Unit IP & Broadband [2] Technology & Business Lab
Ericsson GmbH, Aachen, Germany
{andreas.fasbender,martin.gerdes,sascha.smets}@ericsson.com

Abstract. The Intelligent Transport Systems (ITS) domain is already today the leading vertical industry sector for the adoption of cellular connectivity. In-built and brought-in access to mobile telecommunication networks is a key enabler for advanced safety, navigation and infotainment services.

The growing complexity of the industry ecosystem as well as the diverse service requirements put on the underlying connectivity and service enablement infrastructure demand for open, flexible and scalable end-to-end ITS solutions.

In this paper we present scenarios, solution requirements, service enablers and an end-to-end ITS system architecture, covering in-vehicle and backend components, connectivity solutions and service life-cycle management. While we put specific emphasis on presenting opportunities and challenges relating to cellular ITS solutions, we also point out aspects relating to the required marriage with Dedicated Short Range Communication (DSRC) systems.

Complemented by experiences from projects with industry partners and the research community and by reflections on ongoing efforts in ITS standardization, we conclude that only by a combination of cellular and DSRC networking technologies the full range of consumer and business needs will be addressed.

Keywords: ITS, HSPA, LTE, 802.11p, DSRC, connectivity enablement, flexible subscription management, service enablement, OSGi, virtualization.

1 Introduction

With about 40% of connections or ~40M subscriptions the transport industry is already today the largest vertical market for cellular M2M connectivity [1]. And the adoption of mobile technology in the ITS sector is expected to even accelerate, and double until 2014. The rapid uptake of vehicular connectivity is driven by e.g.

- the demand for increased road safety: ~1.3M people yearly die as a result of a road collision, which is one of three leading causes of death for people aged between 5 and 44 years and costing countries 1-3% of GDP [2];
- the optimal use of road infrastructure to cope with sharply increasing traffic volumes: ~1.2B vehicles are projected globally by 2020, up 73% from 1998 levels [3];

A. Vinel et al. (Eds.): Nets4Cars/Nets4Trains 2012, LNCS 7266, pp. 1–13, 2012.

- macro-economical factors such as the rapid growth of megacities and forecasts that 50% of urban population will not own a car by 2050, driving the demand for co-modal and multi-modal urban transport solutions as well as traffic efficiency;
- the goal to drastically reduce CO2 emissions globally while increasing energy efficiency and the share of renewable energies, in the EU expressed by the 20-20-20 renewable energy directive;
- the global push towards an accelerated introduction of electric vehicles and supporting infrastructure, and similar industrial and political forces aiming to embrace e.g. pay-as-you-drive road pricing and insurance models;
- the need of passengers and commuters to always stay online and enjoy constant access to their personalized Internet service portfolios, such as radio and video streaming, navigation and social networking services.

In this article we summarize selected findings from working on the adoption of cellular networking technologies in the ITS solution domain. Based on collaborations with the automotive industry, we assess requirements and impacts of ITS services on the underlying connectivity and service enablement infrastructure and ecosystem.

We start with sample usage scenarios in Section 2 that drive the adoption of vehicular connectivity, deriving ecosystem impacts and solution requirements. Section 3 introduces an open, flexible and scalable ITS solution architecture and selected core enablers for advanced, mobile telematics and infotainment services. Here we also reflect on past and ongoing R&D work towards combined cellular and short-range ITS solutions. Section 4 summarizes our conclusions and outlook for future work.

2 Scenarios and Requirements

2.1 Example Scenarios

Mr. Martin lives in Belgium, where he buys a brand-new sedan from a German automaker. It is the year 2014, where services such as emergency call (eCall), network-assisted navigation, breakdown assistance and automated maintenance scheduling are included as standard features. Adding to his basic portfolio, Mr. Martin books personal assistance services, advanced RT navigation (including hazard warnings, virtual traffic signs, and travel and parking guides), Internet media streaming services, and guest Internet access via the car-internal WiFi hotspot into his package.

Provisioning

Upon signing of the contract at Mr. Martin's dealer, the car is ordered at the automaker. Once delivered to the dealer, the car is pre-provisioned for the Martins: NFC-enabled car keys are paired with the car for Mr. Martin and his wife to support personalization of e.g. seat height, steering wheel, and preferred media sources.

The communication gateway in the car is auto-provisioned to connect by default to a Belgian operator as home operator. Via a portal Mr. Martin specifies that traffic consumed by his wife, children and two ride-share colleagues will be charged to his car subscription, while traffic generated by other passengers will be charged to them.

Remote Car Management and Assistance

A multitude of sensors constantly monitors essential car functions. After 5.000 km of driving, oil levels are below minimum and the brakes require attention. This information is – via the OEMs Customer Relationship Management portal – delivered to the dealer, who in turn informs Mr. Martin about a required maintenance action. By pressing the assistance button, he is instantly connected to a personal helpdesk at his car vendor, and reserves a booking at the garage.

General car usage information is permanently collected for statistics and OEM quality assurance purposes, but anonymized before being uploaded to the car maker. Personal driving data of Mr. Martin and his wife are also aggregated and forwarded to their insurance company once per week via a secure channel.

Telematics Services

The car also permanently tracks environmental parameters such as speed, acceleration, location, road surface conditions, traffic signs, and weather situation. One day, as Mr. Martin approaches a curve, the navigation system warns him about an accident and a related traffic jam right ahead. He hits the brake, but a notch too hard and the tires block due to the icy road conditions in front of the curve. The system reacts instantly, resulting not only in a safe standstill, but also in a location-augmented "black ice" warning that will now be spread to all vehicles approaching this road stretch.

Infotainment

Prior to a vacation trip to Italy, Mr. Martin has synched the latest media between his DLNA home media server and the car, so that the kids can enjoy the latest movies on the rear-seat entertainment system. Later, his son plays an online multi-party game, while his daughter connects to her social network. While Mrs. Martin prefers listening to online music stored in her personal media bank using her phone, Mr. Martin activates his headset via voice-control and tunes into his favorite Internet radio channel.

Despite crossing several borders, the Martins do not need to worry about roaming costs for the considerable network capacities used during this trip.

2.2 A Highly Complex Ecosystem

These sample scenarios already motivate that the growing adoption of vehicular connectivity has an immense impact on the business foundation and technical complexity of the future ITS ecosystem (Figure 1), with new challenges relating to for example OEM logistics, car sales and after-sales processes, service business models, HMI integration, and telecoms and ITS standardization, to name just a few.

Car OEMs are now adding cellular broadband connectivity as a hygiene feature into their new product lines. At the same time, they increasingly adopt service-based sales models. For many OEMs, more than 50% of income is already today generated after the initial car sale. Emergency call, road pricing or theft tracking services will be mandated in many countries within the next years, implying that road authorities, road tolling consortia, insurances and other authorities will enter the ITS ecosystem. Internet and media Service Providers are also pushing into the established automotive value networks, based on Internet business and deployment models. This also adds significant complexity to the in-car electronics subsystem and HMI, as 'foreign' applications start competing for driver attention and with vehicle resources that have so far lived in a well-shielded IT environment.

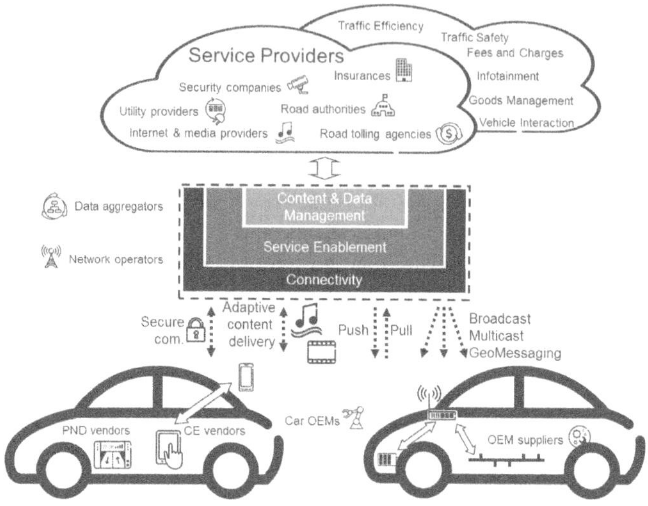

Fig. 1. Growing complexity of the ITS services ecosystem

Backend and in-vehicle communications will be enabled by a number of different solutions, and most operators have reacted by setting up specific M2M business units. Approaches include built-in gateways by OEM suppliers, brought-in (aftermarket) gateways, and the user's mobile phone or PND acting as display and/or routing device, adding gateway, PND and phone manufacturers to the ITS value network.

2.3 Solution Requirements

The future connected vehicle will link different devices to Service Providers from various domains, in many cases via several access networks and operators. Each Service Provider will impose specific requirements for example on bandwidth, real-time constraints, provisioning, data security, and APIs, as described in the following.

Connectivity Enablement

Requirements on connectivity span from guaranteed and instant delivery of life-critical information, to high-latency data aggregation and best-effort Internet bearers. Today's cellular and fixed (data) communication networks deliver and enforce different connectivity packages and types. To fully cater for automotive service demands, innovative support systems are needed that among others provide:

- support for full *IP connectivity*, both over enterprise networks and public Internet;
- *late binding* of network subscriptions to an operator after car delivery, and optional *re-provisioning of subscriptions* in the field during vehicle lifetime;
- efficient *on-demand provisioning and management* for large pools of subscriptions;
- capabilities to create and enforce *tailored*, i.e. service dependent, *connectivity services* with specific QoS and pricing, security, charging and billing rules;

- access-agnostic *connection management*, mediating between backend enablers and available cellular and/or short-range bearers depending on service needs.

Connected Vehicle Platform and Connectors

Car OEMs adopt vendor-specific strategies for integration of online services within the vehicle and for backend connectivity. A generic ITS solution requires that it

- is as far as possible *independent of* the underlying *connectivity hardware and OS*;
- can be *integrated with different in-car networks*, sensor networks, media buses and consumer electronics (e.g., via CAN, MOST, LIN, AutoSAR, BT or WiFi);
- provides support for *Life Cycle Management*, i.e. can be flexibly enhanced after car sale to support both new or modified in-vehicle devices, as well as future services and applications and their upgrades;
- provides *generic connectivity enablers* like e.g. remote provisioning, remote wake-up, and device, service and performance management;
- can *guarantee reliability, security and QoS enforcement* where required.

Service Enablement

Service Enablement refers to generic functions that can be reused to compose domain-specific services. These can be applied across different ITS service ecosystems, and in some cases even across different vertical industries.

Service enablers can be hosted by different players in the value network. Example enablers include user profile and context management, identity federation, positioning, geo-messaging, QoS management, data aggregation and filtering, scheduling, and content transcoding. Centralizing generic functions at a network operator or aggregator leads to reduced capital and operating expenses due to better economies of scale.

A component-based service enablement approach

- supports *deployment across different ecosystem partners* (e.g. MNO, connectivity aggregator, location services provider, service or infrastructure hosting provider);
- provides in-built *scalability, performance, and security*, e.g. through cloud-based deployment and advanced AAA and SLA management;
- adopts *service-oriented architecture* principles and *web programming standards,* supporting simple composition and flexible integration with other service enablers;
- enables *data and user anonymity* where requested, e.g. by using ID masking and separating data management from content management;
- *exposes functionality over APIs*, be it provider internally, via SLA-controlled links (e.g. to OEM or insurance firm), or completely open to other Service Providers.

Content and Data Management

ITS solutions typically also require integration of domain specific enablers, building on the generic service enablement functions described so far (see Figure 2).

Examples of Content & Data Management enablers include real-time road usage data aggregation, data fusion of vehicle battery charging needs for power generation planning, device/content aggregation, transcoding and channel bandwidth management for optimized media delivery, or network-based voice recognition. Such enablers are data and information centric, and tightly integrated with the Business Support Systems (BSS) of the affected Service Provider. They should, among others,

- support *deployment across different ecosystem partners*, (e.g. network operator, data aggregator, cloud hosting provider, road toll authority or media company);
- provide *APIs for flexible integration into external BSS*;
- allow for the *distribution across administrative domains* to enable the federation of information and services.

In this article, we prioritize issues relating to connectivity and service enablement, and will only briefly touch some content and data management related issues.

3 An Open, Flexible and Scalable ITS Solution Architecture

In the following we describe a layered end-to-end solution architecture for ITS services, covering Connected Vehicle Platform (CVP) and corresponding components in the telecommunication and ITS backend (Figure 2).

Connected Vehicle Platform

The *connected vehicle platform* (CVP) addresses the vehicle side (OEM hardware and software) requirements of the proposed ITS solution architecture.

A central role of the CVP is to open up the capabilities of the ICT world to the automotive world to enable advanced ITS services utilizing communication technologies. While the CVP architecture proposed in this paper integrates the communication platform with the application environment, the proposed end-to-end ITS solution architecture does not exclude platform distribution into a Network Access Device (NAD) and one or more in-vehicle application platforms.

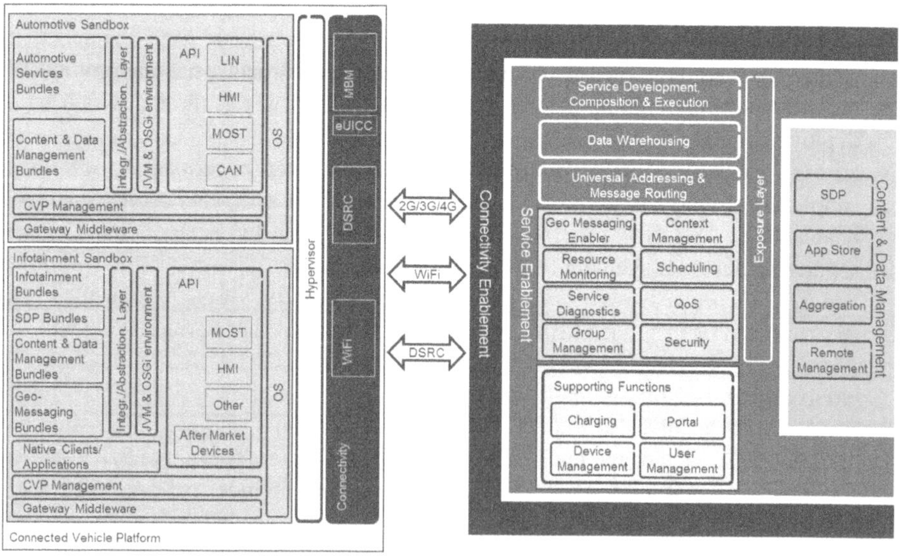

Fig. 2. Generic end-to-end ITS solution architecture

Human Machine Interfaces (HMI) can be realized through applications running on the CVP, inbuilt devices connecting to the CVP, and or portable user devices such as smart phones or advanced PNDs with support from the CVP.

Virtualization of the CVP hardware using a hypervisor has portability and deployment advantages across different application domains: Multiple CVP software platforms including their dedicated OS can be deployed as confined sandboxes on top of the virtualization environment, independent of the underlying hardware platform.

Enabling *access and priority control for independent sandboxes* allows assigning different subscriptions to each application domain. By that the access to security relevant or OEM internal applications can be separated from the more open user applications. For example, safety applications shall not be degraded in their availability, usability and performance by infotainment applications running on the same shared platform. Here, virtualization can ensure that safety-critical processes always get required hardware resources like CPU time and prioritized communication capacity.

Connectivity

Different radio access technologies can be supported through corresponding plug-in modules in the CVP, e.g. cellular 2G/3G/4G, WiFi 802.11p, and DSRC. OS and middleware provide standard APIs towards network applications, native clients, and for service enablers through an integration and abstraction layer. *Remote wake-up functionality* enables the CVP to remain in sleeping mode during parking, consuming almost no power. Via the cellular network, the platform can be woken up on demand from a device management component with corresponding APIs.

Beside utilizing the networking capabilities of the CVP for applications running on the platform itself, the gateway middleware can also provide *connectivity for nomadic devices* carried into the vehicle, such as smart phones, laptops, or tablet PCs, connected via local access networks such as WiFi and Bluetooth. Based on supporting functions such as user management and charging, a *smart bit-pipe* can be established, providing service-aware QoS provisioning and charging for services on nomadic devices. Thereby, billing for vehicle connectivity and for automotive specific services (e.g. safety services) can be separated from billing for user-centric services (e.g. remote content access or media streaming services).

Through implementation work in the CoCarX [12] project and in bilateral engagements we could prove that smart bit-pipe services can be established by either using IMS for service differentiation, or by relaying WiFi-based local authentication via the CVP SIM to the operator subscriber management support systems.

Interworking of Cellular and Ad-hoc Networking Technologies

Today, different wireless access technologies are often still seen to be competing for mass deployment of ITS products and services. Based on results from R&D projects with the automotive industry, we promote an integrated approach, where *cellular and short-range access technologies* complement each other as communication enablers for an even broader range of applicable services. This is also a conclusion from [11]. In the German AKTIV CoCar project [5], High Speed Packet Access (HSPA) networks were shown to be capable of handling a variety of ITS services. Even time-critical safety related services like Road Hazard Warnings can be served. It was proven that the *car-to-network-to-car transmission delays* in a HSPA network are in most cases below 500 ms (Figure 3).

Long Term Evolution (LTE) networks will reduce car-to-car delays even to values below 100 ms in 95% of cases. The infotainment user experience will then be uninterrupted even for HD video. Due to the drastically increased system capacity of LTE much more vehicles can be served at the same time in the same area. The LTE roll-out will soon lead to almost complete broadband coverage in broader parts of the world. Combining HSPA/LTE performance with the wide infrastructure availability, cellular networks allow a straight-forward introduction of ITS services. This is especially needed for overcoming situations where DSRC radio equipped vehicles are not available within the transmission range [11].

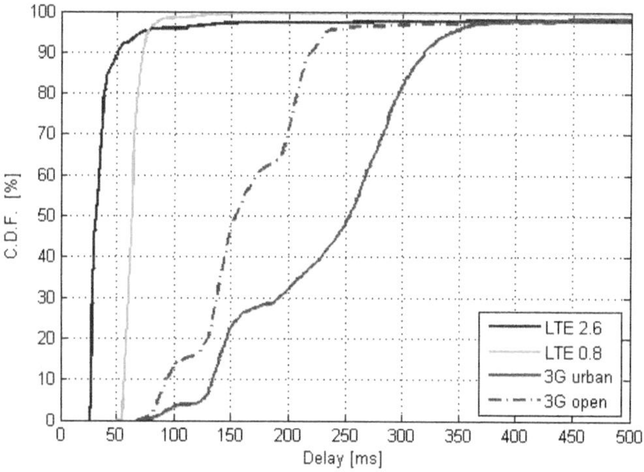

Fig. 3. Car-to-Car delay measurement results from [6]

The CoCarX follow-up project [12] focused on how cellular networks can be combined with ad-hoc solutions such as 802.11p into one *heterogeneous communication system*, and proved that by this combination the strengths of both systems can be leveraged. By using *standardized message formats* (e.g. as currently specified in ETSI Technical Committee Intelligent Transport System (ITS): [13] and [14]), messages can seamlessly be transmitted over the most appropriate channel, depending on the requirements of the service and available communication capacities. While cellular systems offer advantages compared to WLAN-based solutions in terms of coverage and easy migration, ad-hoc networks provide shorter latencies, required e.g. for line-of-sight critical services such as car collision avoidance.

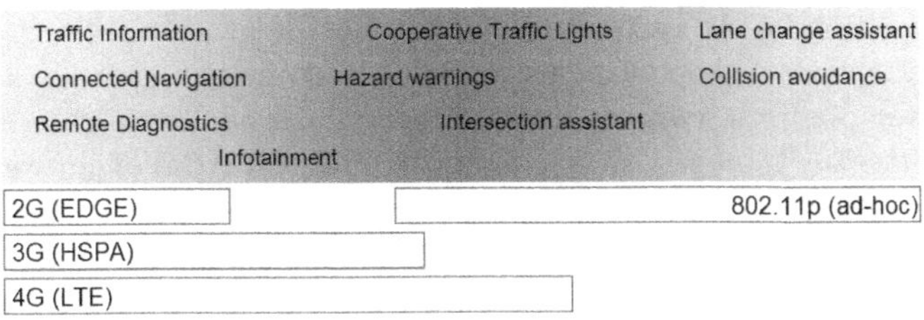

Fig. 4. Complementary network technologies for ITS services

Figure 4 clearly shows that only by complementing communication technologies the full spread of required ITS services can be supported. For an analytical comparison of IEEE 802.11p/WACE and LTE performance the reader is referred to [15].

Provisioning and Subscriber Management

A particular strength of cellular networks is the inherent support for global *connectivity enablement* in globalized car production and sales processes, and for the cross-border mobility of vehicles, including the *flexible provisioning* of network subscriptions. However, in order to enable full cost control for users, vehicle manufacturers, and services providers, traditional data roaming approaches cannot be sustained. Operators are therefore teaming up in international alliances to provide *cross-border M2M services* without roaming fees. Another approach is to establish an intermediary MVNO, i.e. a virtual operator that aggregates connectivity by subcontracting 'real' operators in each serviced country, and who also handles the subscriber management.

If global network connectivity can be offered via a contract with a single M(V)NO, in theory a single global subscription is sufficient to authenticate vehicles to the operator. However, this implies network provider lock-in, a situation that OEMs want to avoid, as the life time of a car series can easily exceed 15 years. Therefore, options to *re-provision subscriptions during vehicle lifetime* via remote management processes are required. Similar flows are imposed e.g. if a car is sold internationally.

When car vendors or service providers work with multiple MNOs across different countries, the target network operator is typically not known at production time, implying that the subscription must be bound to a preferred, typically national MNO after car delivery. This feature is also referred to as *late binding*.

Telecommunication operators and vendors have reacted to these industry requirements. Standardization efforts during recent years in 3GPP [8], GSMA [9] and ETSI [10] have recently resulted in the definition of requirements and solutions for a provisioning chain that supports late binding, re-provisioning and the transfer of subscriptions. To ensure a highly tamper-proof solution, subscription credentials must only be stored in *trusted environments (TRE)*. This can be a standard UICC as used in mobile phones today, or an embedded UICC (eSIM) soldered to the connectivity platform. Operators and players in the automotive domain prefer the latter smart card platform.

In proof-of-concept projects with OEM suppliers, the viability and strong security of flexible subscription provisioning solutions could be validated, using different TRE platforms such as UICC and integrated communication module. Standardization of eSIM-based provisioning solutions is expected to fully conclude within 2012.

CVP Abstraction Layer and end-to-end Service Integration

While native clients and applications are tailor-made for the hardware and OS specifics of a given platform, the concept of an *integration and abstraction layer utilizing Java and OSGi technologies* allows deploying different types of software bundles on top of a flexible, extendable and remotely manageable deployment environment.

South-bound the abstraction layer adapts to *OEM specific hardware APIs* (e.g. for access to CAN bus information), while the north-bound integration layer for OEM-independent service bundles enables efficient development, integration and also updates during vehicle life-time. In the same way, the CVP integration and abstraction layer can be flexibly extended for new hardware devices and their specific APIs.

An example is a *content and data management* bundle, enabling efficient collection, aggregation and communication of vehicular data from multiple car sensors and devices, for example speed, location or diagnostics data.

In the service backend, corresponding service enablers communicate directly with their counterparts on the CVP. These service enablers in turn provide APIs for the development, composition and execution of business solutions, utilizing dedicated development tools and execution environments. Following the example above, the *content management* and *aggregation* components support to make the aggregated data from numerous vehicles available to local services hosted in a *service delivery platform (SDP)* or to 3rd party service providers through *data warehousing*. Optionally, data can be processed and anonymized before exposure.

The enforcement of certain *Quality of Service (QoS) levels* addresses the requirement to prioritize certain types of traffic (e.g. safety-relevant traffic information) and to delay others (e.g. remote diagnostics information). Thereby, QoS levels can be assigned as required per service or SLA, and the system can also be off-loaded during busy hours. This is supported by a generic *scheduling function* in the service enablement layer. As one corner stone of Ericsson's *smart pipe* solution, the scheduler carries out the planning and control when and how to retrieve or send data, based on different input derived from SLAs, service owners and metrics like for example time-of-day, network load, and geographic distribution.

Service Life-Cycle Management and Application Stores

Software and service life-cycle management supports the update of native and Java/OSGi software components from a remote software management service.

Application stores are applications on top of a Service Delivery Platform (SDP) that aggregate and deliver applications, and manage the charging and settlement of transactions. Combining the software and service life cycle management with the application store results in a smart, safe and vendor controlled deployment of services and applications. Based on open web standards and APIs, open car application stores enable developers to reach a new attractive customer base. The application store itself is in general agnostic to a car maker brand, although applications may be optimized for specific human-machine interfaces. This opens up for new revenue streams for all ecosystem partners.

Content can be presented in a tailor-made way to in-vehicle clients, taking capabilities such as screen size, resolution and I/O devices into account. While OEMs will typically take care of the HMI, generic interfaces can be provided ensuring a proper presentation to a wide device range, without forcing the developer to take care. Rules for driver distraction can be enforced in a way that traffic safety is enhanced. Several industry initiatives are working on these issues, e.g. Genivi [16], CE4A [17] and Car Connectivity Consortium [18].

The application store is not only attractive to developers, but to car OEMs as well. Branded application stores can for example provide additional services or limit offers to their customer base, e.g. additional engine power could be bought for a weekend. They can also be hosted by different providers (e.g. network management centers, customer care centers, ad brokers), thereby enabling completely new business models.

Road Hazard Warnings Distributed with GeoMessaging as Example

The distribution of Road Hazard Warnings depends - as various other vehicle applications and services – on accurate geographic information, and data distribution rather to a specific geographical area than to specific individual vehicle. As an efficient method for location-based distribution of information, while assuring privacy for the individual users, we propose GeoMessaging as an addition to the known methods of unicast, broadcast and multicast. As proof of concept of the proposed end-to-end ITS system architecture, we have developed a Road Hazard Warning service, a GeoMessaging enabler and further components for the whole chain from development, via distribution, deployment and provisioning to operation, and presented it at major international conferences.

A content and data management bundle and a GeoMessaging bundle are downloaded and provisioned on top of the CVP integration layer from an OEM specific app store. The first one utilizes information, originating from car sensors and provided through OEM specific APIs, to identify vehicle or road hazards, and takes care for the immediate transmission of the geo-tagged warning information to a hazard information aggregation enabler in the service backend. The latter one forwards GeoMessaging area information to the GeoMessaging enabler in the backend. On reception of certain hazard information, the Road Hazard Warning service, running on the SDP, uses the GeoMessaging enabler to immediately distribute the warning information to all vehicles in the same GeoMessaging area.

Utilizing generic APIs of the CVP integration and abstraction layer, the introduced bundles can be deployed in any corresponding car. The received warning information can be used by in-built HMI devices, or by portable user devices connected locally to the CVP. At the same time the *remote software management* allows software maintenance during the life time of the car.

Our grid based solution can be integrated in today's cellular networks, utilizing particular strengths as wide-area coverage, low latencies and cellular structure to achieve minimal location updates and time-critical delivery of information to a massive amount of clients in specific geographical areas. Simulations show that millions of vehicles can be served in an area matching the size of Germany with a delay of just a few hundred ms [7].

4 Summary and Conclusions

Fueled by consumer benefits, new business opportunities and regulatory demand, vehicle communications is already one of the key connectivity growth drivers in the M2M arena. Cellular networks have proven to be capable of addressing a wide range of ITS services at limited investment risks.

A number of road blocks still have to be removed along the path towards an open, flexible and scalable industry ecosystem. In this paper, we have presented an end-to-end solution approach, spanning across in-vehicle networking and vehicle gateway, connectivity and service enablers, all the way to application stores and APIs for the integration of 3rd party services. We have described selected technology enablers, including remote subscription management, flexible service and software lifecycle management, geomessaging and open service delivery platforms.

Via a number of prototyping projects with key industry partners we could verify that a layered and component-based solution approach supports easy integration, reuse and extendibility of services in vehicles using a single gateway integration layer. Other efforts looked into CVP platform issues and flexible subscription provisioning.

The ITS industry is progressing towards a new telematics and infotainment services ecosystem. The presented solutions will be further explored in consumer, business and technology trials during 2012, while further research will be conducted for example in the area of heterogeneous access technologies and architectures for open data exchange and brokering.

References

1. The Global Wireless M2M Market, Berg Insights (2009)
2. Global Status Report on Road Safety, WHO (2009) ISBN 978-92-4-156384-0
3. OECD environmental outlook to 2030 (2008) ISBN 978-92-64-04048-9
4. Jacobsson, M., Niemegeers, I., Heemstra, S.: Personal Networks: Wireless Networking for Personal Devices. John Wiley & Sons, Chichester (2010)
5. Dietz, U. (ed): CoCar Feasibility Study: Technology, Business and Dissemination, CoCar Consortium, Public Report (2009)
6. Phan, M., Rembarz, R., Sories, S.: A Capacity Analysis for the Transmission of Event- und Cooperative Awareness Messages in LTE Networks. In: ITS World Congress 2011, Orlando, USA (2011)
7. Jodlauk, G., Rembarz, R., Xu, Z.: An Optimized Grid-Based Geocasting Method for Cellular Mobile Networks. In: ITS World Congress 2011, Orlando, USA (2011)
8. Feasibility study on the security aspects of remote provisioning and change of subscription for Machine to Machine (M2M) equipment, 3GPP TR 33.812, Version 9.2.0 (2010)
9. Embedded SIM Task Force Requirements and Use Cases, GSM Association, V1.0 (2011)
10. ETSI, `http://www.etsi.org/`
11. Dressler, F., Kargl, F., Ott, J., Tonguz, O., Wischhof, L.: Research Challenges in Intervehicular Communication, Lessons of the 2010 Dagstuhl Seminar. IEEE Communications Magazine (2011)

12. Fiege, G., Gasser, T., Gehlen, G., Geisler, S., Quix, C., Jodlauk, G., Phan, M.-A., Rembarz, R., Wiecker, M.: CoCarX ITS Services and Communication Architecture, Deliverable D3 (2011)
13. ETSI TS 102 637-3: Intelligent Transport Systems (ITS); Vehicular Communications; Basic Set of Applications; Part 3: Specifications of Decentralized Environmental Notification Basic Service
14. ETSI TS 102 637-2: Intelligent Transport Systems (ITS); Vehicular Communications; Basic Set of Applications; Part 2: Specification of Cooperative Awareness Basic Service
15. Vinel, A.: 3GPP LTE Versus IEEE 802.11p/WAVE: Which Technology is Able to Support Cooperative Vehicular Safety Applications? IEEE Wireless Communications Letters (2012) doi:10.1109/WCL.2012.12.120073
16. GENIVI Alliance, `http://www.genivi.org`
17. CE4A - Consumer Electronics for Automotive, `http://www.ce4a.org`
18. Car Connectivity Consortium, `http://www.terminalmode.org`

Effective and Efficient Security Policy Engines for Automotive On-Board Networks

Muhammad Sabir Idrees and Yves Roudier

EURECOM
{muhammad-sabir.idrees,yves.roudier}@eurecom.fr

Abstract. The configuration of security mechanisms in automotive on-board networks makes it necessary to define and deploy adapted security policies. This paper discusses how to design policy engines that implement an effective enforcement in such architectures despite the complexity of the protocol stacks of on-board electronic control units. It also evaluates how policies expressed in XACML can be adapted to the automotive environment efficiency requirements despite the limited computational power of those units and network bandwidth limitations.

Keywords: Security Policy, XACML, ASN.1, On-Board Policy Engine.

1 Introduction

The changing requirements of automation have triggered a tremendous evolution in automotive on-board systems, making vehicular communication a central feature of modern vehicles. Vehicular on-board networks have long been designed with safety rather than security concerns. The security of bus systems was, to a great extent, achieved through obscurity. Recognized security solutions must now be used in order to mitigate the attacks that recent studies on those architectures have shown to be possible [16] or [25], [14], [33]. Secure communication is a basic need in such architectures and should at least ensure some form of authentication to participants.

Research we conducted in the past 3 years in the European funded project EVITA [21] has been specifically addressing those requirements and appropriate solutions for vehicular on-board networks. This project features a comprehensive security architecture solution covering all aspects of on-board communication (data protection, secure communication, secure and tamper proof execution platform for applications). The security mechanisms that we have defined in this project are in particular coordinated by a security policy aiming at both networking and cryptographic material security. Due to the extended lifetime of vehicles and the need to service them and therefore replace ECUs, it is mandatory that this policy might be updated. We focus in this paper on the networking security policy and discuss how it can be implemented efficiently.

We have approached this problem through the combination of cryptographic protocols and trusted computing, the latter ensuring the binding of the former

A. Vinel et al. (Eds.): Nets4Cars/Nets4Trains 2012, LNCS 7266, pp. 14–26, 2012.

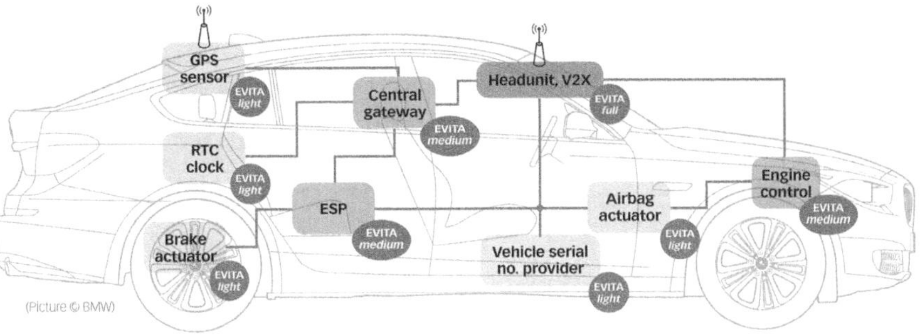

Fig. 1. Automotive on-board network architecture

to a known platform. This is achieved by associating a hardware security module (or HSM) to every ECU, as illustrated in Figure 1. The security policy to be applied in a vehicle is the combination of an invariant policy for the usage control of cryptographic credentials of electronic control units, and a flexible networking security policy. The credential usage control policy is enforced by the HSM [27] and possibly through the virtualization of the ECUs if applications on the same ECU have to be segregated. In contrast, the networking security policy is enforced by all network elements. Those elements first consist in the protocol stacks implemented at the different ECUs: those stacks implement secure protocols for transport, diagnosis, firmware update, key distribution (see for instance [28]), policy distribution, or even system boot; they have to be linked with the keys initially bound to ECUs by the HSM, which provides trusted authentication as the building block for all those protocols. The interconnection gateways between the different communication buses of the vehicle, or the V2X communication units, or even the tools used to diagnose and service vehicles constitute other network elements where policy enforcement can be done, based on trusted authentication or other security mechanisms like traffic filtering or secure logging.

The remainder of this paper is organized as follows: Section 2 outlines the security policy enforcement architecture, and in what way the security policy supports the flexible deployment and enforcement of the networking security policy. Section 3 discusses how XACML is being used to define the flexible part of such a security policy. Section 4 presents and analyzes performance figures for our policy engine. Section 5 reviews the capabilities of existing access control frameworks in comparison with our own system, with respect to the policy interpretation and enforcement. Finally, the paper summarizes the results that we achieved regarding on-board access control.

2 Security Policy Enforcement Architecture

We first describe the architecture that was adopted in [32] to enforce modular access control. The policy engine, which can be flexibly updated, is deployed

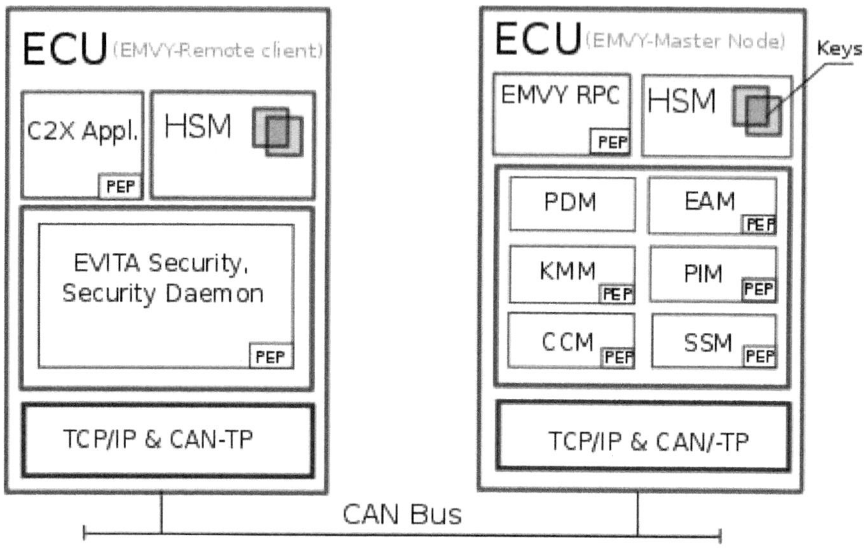

Fig. 2. PDM and PEP deployment

within the automotive system environment. The policy that it enforces is defined as a mandatory policy for all communication in the vehicle at a backend system and is transferred to the vehicle. The policy engine is composed of two main components i) A Policy Decision Point (PDP), following the terminology used in XACML, and ii) Policy Enforcement Points (PEPs). Within the scope of EVITA, both components are implemented as a proof of concept within the EMVY middleware framework defined in [6]. EMVY uses a distributed master-client architecture and provides component based templates for introducing security relevant mechanisms for securing communication between ECUs within a vehicle. The PDP is implemented by a component termed the Policy Decision Module (PDM). PEPs are instead implemented at various components of the middleware, as described below. The policy decision module is used as a centralized module accessible from different domains and application as shown in figure 2.

2.1 Policy Enforcement Points

Policy enforcement points (PEP) have to be deployed together with a PDP. PEPs handle access, or communication, or boot sequence validation requests for instance and have to enforce policy based decisions. Access control in vehicular embedded IT infrastructures has for instance traditionally been developed using a single enforcement point. However, in our fine grained architecture, policy enforcement has to rely on a series of such PEPs at different parts of the communication stacks and middleware, as well as at application level. For instance,

Table 1. Software Security Modules and Policy Enforcement Points (PEPs)

Security Module	**Module Description**	**Policy Enforcement Point (PEP)**
Communication Control Module (**CCM**)	Handles all communication functionality between EMVY nodes and internal modules.	enforce the filtering policies which act as firewall. Rule enforcement will be based mainly on transport-layer parameters but also on application-layer information.
Entity Authentication Module (**EAM**)	provides authentication, single sign-on/ sign-out security services.	enforce the policy based login and authentication services (e.g., password, smartcard).
Key Management Module (**KMM**)	Distribution of fresh cryptographic key material among clients.	enforce the key generation and group communication security rules.
Platform Integrity Module (**PIM**)	Enhancement of platform security by supplying runtime platform integrity	rule enforcement regarding valid boot integrity measurements.
Secure Storage Module (**SSM**)	Cryptographic storage operations for applications.	enforce the secure storage/access of data security rules (i.e. encrypt the storage device or, can encrypt data objects individually).
Security Watch-Dog (**SWD**)	Intrusion detection and management services.	policy enforcement rely on a set of rules consists of a attack pattern and an action.

deciding on access control for some RPC request requires first allowing network traffic between the communicating parties, then distributing the proper session keys through the key distribution mechanism, then authenticating the communication channel using these keys, then finally making sure that the RPC request can be performed by the entity authenticated at the transport layer. A list of the enforcement points deployed in the car is presented in Table 1.

2.2 Handling Policy Decisions

The policy decision point decides based on security policies whether or not access to a particular resource is granted. The decision of the PDP is then enforced by the policy enforcement point (PEP) which drops a message, forwards a message or modifies a message (e.g., encrypts the message with the cryptographic key of the destination ECU) according to the security policy. A policy decision point usually acts autonomously in its domain where he is assigned and makes decisions in response to an authorization request. By default, a PEP actively queries the PDP for every decision. For instance, an EMVY remote client application (C2X application in figure 2) queries the PDM at the EMVY Master Node for each authorization request. However, a PEP could be pre-configured by the PDM (autonomous PEP). Such an autonomous PEP would act as an ancillary PDP, mostly based on static or cached security decisions, and possibly parameterized decisions depending on security information that are available locally. For instance, during the secure bootstrapping phase, the SSM-PEP acts as an autonomous PEP and decides based on its local policy whether the PDP is allowed to load security policies.

3 Security Policy Expression

As we explained, the onboard network policy has to describe how to configure very different security mechanisms. After exploring several alternatives including drafting our own policy language, we decided on building upon the XACML [17] access control language, obviously for expressing the access control rules of our security policy, such as the definition of secure communication groups and related authorizations at the RPC level, but as well as a more general policy language. XACML provides a flexible and modular way to define and enforce policies, and its decision/enforcement model fits well in distributed environments, even for the on-board embedded system of a vehicle. XACML provides an interchangeable policy format, support for the fine grained description of resources, can describe conditional rights, supports policy combination and conflict resolution. Another important aspect regarding the choice of XACML as our policy language was its independence from a specific implementation and the large number of tools for writing and analyzing any policy. In the case of specific configurations, XACML is flexible enough to represent different security profiles. For instance, the XACML `Policy` element can be used in order to encapsulate complex firewall rules comprising multiple attributes, the source IP address/port numbers being specified in the XACML `Subject` element for instance, and the destination IP address/port number being mapped to the XACML `Resource` element. In a similar way, the XACML `Rule` element can be used to represent distinct firewall rules.

However, due to the embedded nature of the on-board system and its functional and non-functional constraints [18], [15], it is not feasible to transmit, process, or store XML-based policies in the car. For instance, for the ubiquitous CAN bus, which is operated at around 500kbit/s and offers 8 bytes of data payload per packet, verbose formats like XML would constitute a hardly justifiable increase of the bus load. To cope with the above-mentioned limitations, we defined a binary-based security policy language that consumes less bandwidth, is fast to process, and requires less memory (see Section 4 for a performance analysis). We called this representation the Policy decision module Native Language (PNL). The purpose of the PNL is not to define yet another access control policy language but rather to provide an alternative interchangeable format for XACML policies, that can be used where performance is an issue. We built PNL on ASN.1 standards [31]. These standards are adopted in a wide range of application domains, as in aviation systems for traffic control, mobile networks, network management, secure emails, fast web services, etc., [31]. PNL makes use of these standards, describes a serialized representation of XACML policies in binary format, and ensures that the XACML structure is preserved during serialization. In order to do so, a XACML schema is mapped into a corresponding ASN.1 definition (see listing 1.1). This mapping is based on the ITU-T X.694 standard (Mapping from XML Schemas to ASN.1 modules) [30].

```
XACML DEFINITIONS AUTOMATIC TAGS ::= BEGIN
/* XACML Policy Defintion */
PolicyType ::= SEQUENCE {
  ...
  ruleCombiningAlgId UTF8String,
  target Target,
  choice-list SEQUENCE OF CHOICE {
  ...
  rule Rule
 } OPTIONAL,
 obligations Obligations OPTIONAL
}
Policy ::= PolicyType
/* XACML Target Definition */
TargetType ::= SEQUENCE {
  subjects  Subjects OPTIONAL,
  resources  Resources OPTIONAL,
  actions   Actions OPTIONAL,
  environments  Environments OPTIONAL
}
Target ::= TargetType
...
/* XACML Rule Defintion */
RuleType ::= SEQUENCE {
  effect  EffectType,
  ruleId  UTF8String,
  description  Description OPTIONAL,
  target  Target OPTIONAL,
  condition Condition OPTIONAL
}
Rule ::= Rule
...

END
```

Listing 1.1. Excerpt of a Policy decision module Native Language (PNL) based on ASN.1 Definition. An excerpt of serialized policy using listing 1.1 is presented in Fig.3.

We have implemented the XACML to PNL mapping engine. In our architecture, this encoder resides at the OEM's backend system. The mapping engine is responsible for serializing security policies into the PNL format and then transmitting it to the vehicle. During the serialization process each security policy is verified against a XACML schema. Upon a successful validation, the security policy is serialized into a specific ASN.1 encoding scheme. Due to constraints from other components (e.g., low level drivers, HSM interfaces, etc.) which employ the ASN.1 DER encoding [29], the security policies are also serialized using DER encoding rules. However, a more efficient binary encoding such as Packed Encoding Rules (PER) can be used if needed. We anticipate that using PER encoding scheme would further enhance performance and latency results.

3.1 Security Policy Configuration

From the on-board PNL configuration perspective, our high-level goal is to auto-configure security policies and appoint access rules from the vehicle ignition stage. This implies that security policy are deserialized during secure bootstrapping phase, while also recognizing an explicit configuration procedures, after secure bootstrapping phase, in order to load new/update policies (i.e., installation of new application with its policy set or update existing policy). Note that the new policy set or security rules which may have an impact on the basic safety of the vehicle are always configured during next vehicle start cycle. Adding/updating safety critical rules while vehicle is running may cause safety critical problems, depending on the security policy responsible for.
In our implementation stack, the bootstrapping protocols are defined to ensure the secure initialization of all security modules and components (see figure 2). On a certain point of the boot strapping procedure, the boot chain send an initialization call to the PDP(s) to load all PNL based security policies (i.e., group communication policy) from its policy database. Note that these security policies are stored in the Secure Storage Module (SSM), which enforces confidentiality, integrity, authenticity, and freshness mechanisms. Thus, require proper authentication and access rights to access these policies. Since during boot strapping process, SSM cannot ask PDP for a decision when PDP is opens/reads from its policy database (which is also stored via SSM), it has to make autonomous decisions. A detail description of autonomous decisions (autonomous PEP) is discussed in section 2.1. On successful verification of access rights, the SSM allow the PDP(s) to read policy set from its policy database. Regarding the integrity of policies, we rely on the security solutions enforced by SSM. However for the policy validation attempt in our prototype implementation, we enforced, that

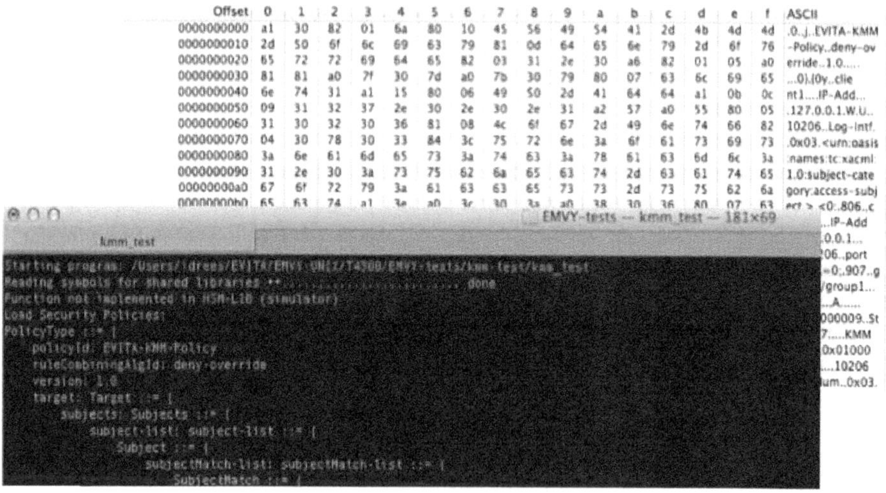

Fig. 3. PDM: On-board policy deserialization and configuration

a vehicle will only be started in case of successful configuration of all PDP/autonomous PEP(s) with respective security policies and the vehicle will shutdown momentarily otherwise.

4 Performance Analysis

We have analyzed the performance of the PNL as well as the memory consumption of different policies by varying the number of elements and attributes used. Performance has a special importance with respect to the user experience in automotive environments, where the time to load and configure security policies, and to assess authorization request and response time is a critical issue when the driver waits for his vehicle to start. The deserialization and configuration of authorization/security policies must be performed before receiving any request from a PEP. Hence, the configuration of these security policy is significantly contributing to the overall responsiveness of the policy decision and enforcement at startup time.

4.1 Performance Analysis: Technical Approach

In order to evaluate our results against other XML based access control policies, we are comparing our results with different XML parsers. There have been numerous benchmark studies towards the evaluation of XML parsers [3], [8], [13], [2], [34], [26]. These benchmarks cover several aspects of XML parsing such as performance, schema validation, DOM manipulation, XML security, etc. We used these benchmarks in order to select an agile XML parser implemented in the same language as our own (C/C++), and to run a comparison with our ASN.1 encoded policies. Several lightweight C/C++ XML libraries [3], [2], [1] have been developed for low power devices, with fast parsing capabilities. For instance, the pugixml library enables extremely fast, convenient, and memory-efficient XML document processing. It consists of a DOM-like interface with rich traversal and modification capabilities. However, since pugixml has a DOM parser, it cannot process XML documents that do not fit in memory; also the parser is not a schema validating parser [3], which is a mandatory requirement in our case, for obvious security as well as safety reasons, such as to prevent software compromises.

4.2 Experimental Setup and Results

Based on results of these benchmarks and these specific constraints, we decided to evaluate our results with the Gnome XML toolkit (libxml) [4]. According to the XMLBench Project [8], libxml is the fastest toolkit that has a rich enough set of features. We have generated several XACML policies with various sizes using the UMU XACML editor. The policy tickets are than transformed into a corresponding PNL description, as described in section 3. We have set up a testing environment in order to compare the scalability of our parser with libxml. All presented performance results were obtained using a 64 bit Mac OS X 10.7.2 on a MacBook

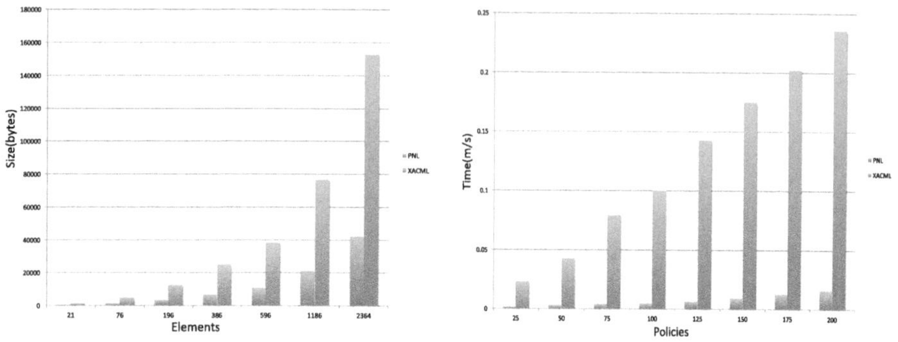

(a). Comparison of XML and PNL security policies, with different elements and attributes, in terms of size (bytes) of policy data

(b). Speed up analysis of about 25-200 XML and PNL security policies, where single security policy is constitute of 21 XACML Policy Elements.

Fig. 4. Size of data and increase in speed up factor

Pro with 8GB of RAM and a 2.8GHz Intel Core i7 processor. All tests were run in a single user mode without any system services running. We followed the assumptions outlined in [8]: for instance, the time spent to initialize the toolkits is not counted in these results, in order to compare our results with this benchmark. The measurements consist in several latency results which show that the parsing of a PNL encoded policy is lighter and enables a much faster parsing and configuration of security policies. Figure 4.a compares the parsing time with different policy sizes, in which the PNL encoded policy parsing is approximately a third of the XML policy. For 2364 elements (2898 attributes) in a single XACML policy, the PNL encoded policy amounts to 42,368 bytes and the XML encoded one to 152,817 bytes. In the eight tests presented on Figure 4.b, we parsed a single XML policy (21 Elements, 22 Attributes) up to 200 times to understand the scalability. This resulted in parsing times of 0.25ms. In contrast, deserializing using the PNL encoded security policy 200 times takes approximately 0.015ms. Parsing with the PNL encoding is thus about 16 times faster that with an XML based encoding. If we scale up the above presented comparison results for modern ECUs which might only contain 32 bits, 40-MHz to 60 MHz processor [11], the parsing time for 200 PNL policies in an average ECU takes up to 1.05ms. Whereas, the same XML based policies requires approximately 17.5ms. Furthermore, a link to the maximum computational delay to provide safety applications (also related to the maximum number of signature verifications per second) might also highlight the need of such fast and scalable binary based policy language.

5 Related Work

There has been a quite remarkable progress in the area of access control architecture for automotive networks [10], [7], [20], but they appear to have succeeded mostly in terms of requirement specification or have been only concerned with security policies for protecting V2X communications. Gerlach et al. [12] present a C2C communication solution integrating several previous proposals [23], [24], [19] for secure vehicular communications [7]. These proposals consider the car mostly as a single entity, communicating with other cars using secure protocols and thus essentially aim at communication specific security policies enforced at the Communication Unit of the vehicle. Our approach in contrast treats both the expression of V2X and intra-vehicle security policies uniformly. The EASIS project [10] defines a central gateway, a sort of firewall, that is configured so that it denies all data traffic from the external interfaces (e.g. C2C/C2I or Telematics) as a default. Unfortunately, like [7] this proposal is also limited to V2X security policy enforcement and not accompanied by any further analysis of the particular requirements/limitation of an in-vehicle architecture with respect to security policies. Zrelli et al. [35] proposed a security framework for the vehicular communication infrastructure implementing access control at both the data link layer and the network layer. However, the proposed solution is solely based on a central policy decision and enforcement module. A single failure in this module may compromise the overall security of the on-board network. Furthermore, this solution is obviously only handling V2X security policy enforcement at the gateway level.

For in-vehicle architectures, numerous authors mention the need for an on-board access control architecture [16], [22], [5]. However, very few solutions have been proposed. A recent security analysis [16] has shown that the risk of attacks on vehicle on-board systems is not anymore of theoretic nature. It depicts several scenarios where access control is either weak or simply not considered, like the firmware update process, which may compromise the overall security of the on-board network, yet no security architecture is described in this work. In [5], a set of cryptographic protocols are discussed to support vehicular use-cases. However, regarding access control, this intra-vehicle security tool box is also limited to the only specification of an API, without any detail about the policy decision engine, practical matters regarding the enforcement architecture, nor implementation perspectives.

Chutorash et al. [9] propose an approach for integrating firewalls in a vehicle communication bus. Firewalls are integrated between application software and between vehicle components. In their approach, filtering rules are applied only on user's request, and commands sent from the HMI, preventing unauthorized access to vehicle components. We see that the practicality of this approach is largely limited by the fact that: `i)` rule enforcement is limited to firewall rules and more specifically only to user commands. However, in an automotive system, different entities (i.e., security modules as discussed in section 2) are themselves requesters `ii)` rules are statically defined and remain the same over the vehicle lifetime, and `iii)` the constraints of embedded vehicular networks regarding notably the policy transmission, processing, or storage and the practical implementation of the proposed approach, are again left out. The OVERSEE Project [22] aims among other objectives at in-vehicle firewall configuration and application level access control using XML based configuration rules. As of now, this project has just begun and no result is available, thus we cannot evaluate the practicality of this approach. However, according to our experience, the performance of verbose formats like XML in a constrained environment has to be closely watched.

6 Conclusion

We have exposed in this paper in what respect the complexity of automotive on-board network architectures and their evolution involve a complex expression of the security policy. Today's automobiles are a perfect example of a system whose security relies on the combination of many different enforcement points in the vehicle on-board network and even in every electronic control unit's communication stack. The role of a security policy in this context is to link the trusted computing base and the trusted credentials it stores with enforcement mechanisms, as well as to connect enforcement mechanisms together. We are taking advantage of the extensibility of XACML subjects to associate attributes, and in particular the means to perform a trusted authentication of electronic control units: this mechanism is at the core of the EVITA approach. The performance of the policy parsing is also very important in a vehicle. We described how the XACML policy can be encoded in ASN.1 in order to make it fit better the resource constrained environment of a vehicle. The policy engine described in this paper was finally deployed in the EVITA project demonstrator - two cars equipped with the EVITA HSMs and software framework - and was successfully used for network filtering, configuring secure group communication, and RPC level access control.

Acknowledgment. This work has been carried out in the EVITA (E-safety Vehicle Intrusion proTected Applications) project, funded by the European Commission within the Seventh Framework Programme for research and technological development.

References

1. Arabica XML and HTML Processing Toolkit, `http://www.jezuk.co.uk/cgi-bin/view/arabica`
2. Asm-Xml Benchmark, `http://tibleiz.net/asm-xml/benchmark.html`
3. Pugixml Benchmark, `http://pugixml.org/benchmark/`
4. The XML C Parser and toolkit of Gnome libxml, `http://www.xmlsoft.org`
5. Bar-El, H.: Intra-Vehicle Information Security Framework (September 2009)
6. BMW. EMVY: The Embedded Vehicular IT Security Construction Kit, Basic Concept (June 2009)
7. C2C-CC. Car2Car Communication Consortium, `http://www.car-to-car.org/`
8. Chilingaryan, S.: The XMLBench Project: Comparison of Fast, Multi-platform XML Libraries, pp. 21–34. Springer, Heidelberg (2009)

9. Chutorash, R.J.: Firewall for vehicle communication bus. In: International Patent Classification 7, WO/2000/009363, PCT/US1999/017852. European Patent Office (February 2000)
10. EASSIS. Security and firewall concepts for gateways. Technical Report Deliverable D1.2-12, EASIS-Project (2006)
11. Freescale. Mpc565 reference manual. Technical report, Freescale Semiconductor (2005)
12. Gerlach, M.,Leinmüller, T., Goldacker, G., Festag, A., Harsch, C.: Security architecture for vehicular communication. In: WIT 2005 (2005)
13. Cheng Haw, S., Krishna Rao, G.S.V.R.: A comparative study and benchmarking on xml parsers. In: The 9th International Conference on Advanced Communication Technology, vol. 1, pp. 321–325 (February 2007)
14. Hoppe, T., Kiltz, S., Dittmann, J.: Automotive IT-Security as a Challenge: Basic Attacks from the Black Box Perspective on the Example of Privacy Threats. In: Buth, B., Rabe, G., Seyfarth, T. (eds.) SAFECOMP 2009. LNCS, vol. 5775, pp. 145–158. Springer, Heidelberg (2009)
15. Kelling, E., Friedewald, M., Leimbach, T., Menzel, M., Säger, P., Seudié, H., Weyl, B.: Specification and evaluation of e-security relevant use cases. Technical Report Deliverable D2.1, EVITA Project (2009)
16. Koscher, K., Czeskis, A., Roesner, F., Patel, S., Kohno, T., Checkoway, S., McCoy, D., Kantor, B., Anderson, D., Shacham, H., Savage, S.: Experimental security analysis of a modern automobile. In: 2010 IEEE Symposium on Security and Privacy (SP), pp. 447–462 (May 2010)
17. Moses, T.: eXtensible access control markup language TC v2.0 (XACML) (February 2005)
18. Navet, N.: Automotive communication systems: from dependability to security. In: 1st Seminar on Vehicular Communications and Applications (VCA 2011), Luxembourg (May 2011)
19. Papadimitratos, P.: Securing vehicular communications - assumptions, requirements, and principles. In: Workshop on Embedded Security in Cars, ESCAR (2006)
20. CVIS Project, Cooperative vehicle infrastructure systems, `http://www.cvisproject.org/`
21. EVITA Project. E-safety vehicle intrusion protected applications, `http://www.evita-project.org`
22. OVESEE Project. Open vehicular secure platform, `https://www.oversee-project.com/`
23. Raya, M., Papadimitratos, P., Hubaux, J.-P.: Securing vehicular communications. IEEE Wireless Communications Magazine 13, 8–15 (2006)
24. Raya, M., Jungels, D., Papadimitratos, P., Aad, I., Hubaux, J.-P.: Certificate revocation in vehicular networks. Technical report (2006)
25. Rouf, I., Miller, R., Mustafa, H., Taylor, T., Oh, S., Xu, W., Gruteser, M., Trappe, W., Seskar, I.: Security and Privacy Vulnerabilities of In-Car Wireless Networks: A Tire Pressure Monitoring System Case Study. In: Proceedings of the 19th USENIX Security Symposium, Washington, DC (August 2010)
26. Schmidt, A., Waas, F., Kersten, M., Carey, M.J., Manolescu, I., Busse, R.: Xmark: A benchmark for xml data management. In: VLDB, pp. 974–985 (2002)
27. Schweppe, H., Weyl, B., Roudier, Y., Sabir Idrees, M., Gendrullis, T., Wolf, M.: Securing car2X applications with effective hardware software codesign for vehicular on-board networks. In: VDI Automotive Security 27. VW-Gemeinschaftstagung Automotive Security, VDI Bericht 2131, Berlin, Germany (October 2011)

28. Schweppe, H., Roudier, Y., Weyl, B., Apvrille, L., Scheuermann, D.: Car2x communication: securing the last meter - a cost-effective approach for ensuring trust in car2x applications using in-vehicle symmetric cryptography. In: 4th IEEE International Symposium on Wireless Vehicular Communications, WIVEC 2011, San Francisco, CA, United States (September 2011)
29. International Telecommunication Union. Information Technology - ASN.1 encoding rules: Specification of Basic Encoding Rules (BER), Canonical Encoding Rules (CER) and Distinguished Encoding Rules (DER), ITU-T Recommendation X.690. Technical report, ITU-T (2002)
30. International Telecommunication Union. Information Technology - ASN.1 encoding rules: Mapping W3C XML schema definitions into ASN.1, ITU-T Recommendation X.694. Technical report, ITU-T (2004)
31. International Telecommunication Union. Information Technology - ASN.1 encoding rules: Abstract Syntax Notation one (ASN.1): Specification of basic notation, ITU-T Recommendation X.680. Technical report, ITU-T (2008)
32. Weyl, B., Wolf, M., Zweers, F., Gendrullis, T., Sabir Idrees, M., Roudier, Y., Schweppe, H., Platzdasch, H., Khayari, R.E., Henniger, O., Scheuermann, D., Fuchsa, A., Apvrille, L., Pedroza, G., Seudie, H., Shokrollahi, J., Keil, A.: Secure On-board Architecture Specification. Technical Report Deliverable D3.2, EVITA Project (2010)
33. Wolf, M., Weimerskirch, A., Paar, C., Bluetooth, M.: Security in automotive bus systems. In: Proceedings of the Workshop on Embedded Security in Cars, ESCAR 2004 (2004)
34. Wu, Y., Zhang, Q., Yu, Z., Li, J.: A hybrid parallel processing for xml parsing and schema validation. In: Proceedings of Balisage: The Markup Conference 2008, Montréal, Canada, August 12-15. Balisage Series on Markup Technologies, vol. 1 (2008)
35. Zrelli, S., Miyaji, A., Shinoda, Y., Ernst, T.: Security and access control for vehicular communications. In: Proceedings of the 2008 IEEE International Conference on Wireless & Mobile Computing, Networking & Communication, pp. 561–566. IEEE Computer Society, Washington, DC (2008)

Security Solutions for Highly Dynamic Car2X Networks in the KoFAS Initiative

Axel Sikora

University of Applied Sciences Offenburg, Badstrasse 24, 77652 Offenburg, Germany
axel.sikora@hs-offenburg.de

Abstract. The provisioning of security for highly dynamic wireless networks, as for Car2X applications is still a major topic, as very specific requirements have to be solved. Those include a perfect privacy level and advanced real-time behavior, and the necessity to work with a public infrastructure (PKI) to support secure authentication.

This contribution analyzes these requirements, discusses the existing approaches, performs a gap analysis and elaborates on proposals to fill these gaps. It describes work in progress within the KoFAS-initiative for the development of a cooperative pedestrian protection system (CPPS).

Keywords: car2x networks, security, privacy, PKI.

1 Introduction

In general, precise perception of the vehicle's environment is an essential requirement for all preventive vehicle safety systems. For example, today's advanced driver assistance systems (ADAS) are based on on-board environment perception sensors.

Cooperative systems may be used alone or in conjunction with image-based systems, like e.g. [3] [1]. The technology can also be used for an omnidirectional perception system for pre-crash safety and as sensor support for intersection safety. Those systems are designed around or based on car2x-communication technologies, such as IEEE802.11p [4], the "IEEE 1609 Family of Standards for Wireless Access in Vehicular Environment (WAVE)" [5]or similar activities in Europe [6], where the higher layers are defined by the Car2Car Communication Consortium [7].

One of these extensions, are developed within the research project Ko-TAG, which is part of the research initiative Ko-FAS (Kooperative Sensorik und kooperative Perzeption für die Präventive Sicherheit im Straßenverkehr, [8]), initiated in 2009. Here, a solution has been proposed, which enriches the current IEEE802.11p protocols, enables it for additional application scenarios, and has been designed to work as an add-on subsystem, which may be seamlessly integrated into IEEE802.11p-based car-to-X-communication. The proposed subsystem protocol combines communication and localization [9].

Various approaches have been developed in order to answer the needs for network security and privacy of these systems. Two actual projects should be mentioned. The

A. Vinel et al. (Eds.): Nets4Cars/Nets4Trains 2012, LNCS 7266, pp. 27–39, 2012.

goal of PRECIOSA (Privacy Enabled Capability in Co-operative Systems and Safety Applications) is to demonstrate that co-operative systems can comply with future privacy regulations by demonstrating that an example application can be endowed with technologies for suitable privacy protection of the location related data of individuals [10]. PRESERVE (Preparing Secure Vehicle-to-X Communication Systems) [11] contributes to the security and privacy of future vehicle-to-vehicle and vehicle-to-infrastructure communication systems by addressing critical issues like performance, scalability, and deployability of V2X security systems.

This contribution analyses the special requirements for this extension (cf. ch. 2), presents existing approaches (cf. ch. 4), performs a gap analysis and elaborates on proposals to fill these gaps (cf. cg. 5).

2 Ko-FAS System Overview

The Ko-TAG project concentrates on the development of a cooperative sensor network between vehicles and vulnerable road users (VRU). Its objective is to add value to existing passive driver assistance systems such as radar- or camera-based systems. At the one hand, this requires to enable the localization of foreign targets even when losing line of sight or in scenarios with multiple objects requiring prioritization. One the other hand the cooperative sensor network provides valuable additional information about the localized target such as its movement parameters or even entire movement patterns that support the accident prediction algorithms at the higher layers. Since those networks in the Car2X environment are highly dynamic the actual topology can change within short time periods regarding to the number of network participants and the reaction time of the system the different scenarios are handled by an underlying net-work communication protocol allowing a prioritization of connected devices and therefore affect the measurement update rate.

Within the Ko-TAG-architecture, it is anticipated that vehicles are equipped with an On-Board Unit (OBU), which communicates with a SafeTAG (ST). This ST can be given to any secondary partner. In its original purpose, a VRU is equipped with a ST. In addition, if the ST is given to another vehicle, a significant extension of the project scope with regard to omnidirectional safety will be achieved. The integration of infrastructure, i.e. at intersections, could further improve the functionality and the accuracy of the system. This intersection use case is currently addressed in a neighbor project (Ko-PER) and can eventually be extended by the Ko-TAG communication and localization. Fig. 1 shows the potential links between traffic objects.

The cooperative sensor system developed in the Ko-TAG project consists of On-Board units (OBU) and transponders called SafeTAGs (ST). The sensor concept allows for a selective localization of STs and a selective communication link between localization units and STs for an efficient use of the available bandwidth (link (2) in Fig. 2). Furthermore there is a lightly modified version of the ST that can be connected to an OBU via a TCP/IP link to provide omnidirectional safety amongst vehicles (link (3) in Fig. 2).

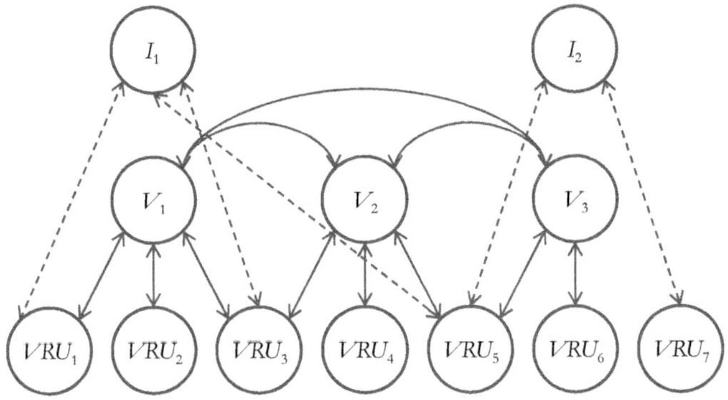

Fig. 1. Supported topologies within the Ko-TAG project; dotted lines show the potential additional links within the overall Ko-FAS approach

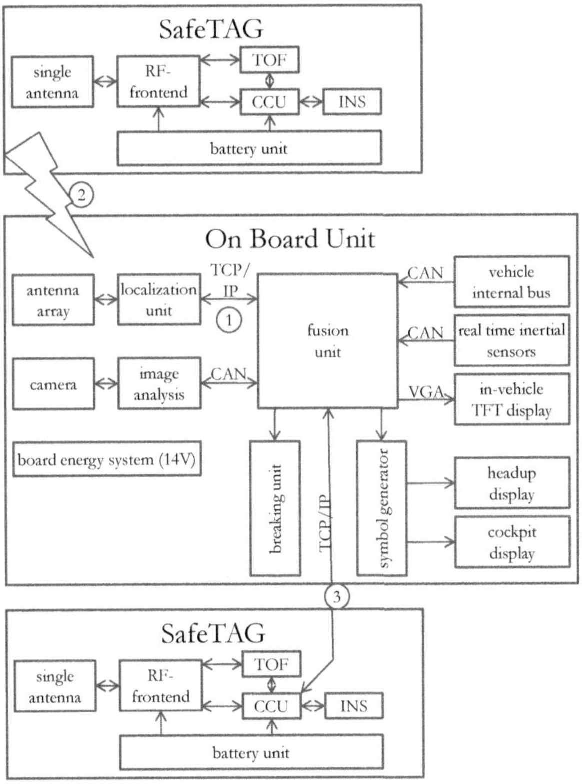

Fig. 2. System Architecture of the complete Ko-TAG system

As of writing this contribution, the project is still on its way – also with many functional challenges. However, the project should pave the way also towards a secure operation of the communication links. Therefore, this contribution describes the current level of discussion within the project – and will report later on its developments.

3 Requirements

3.1 Security Objectives

Literature mainly describes six legacy security objectives [13] [1], which will be denominated as parameters within the course of this document. In [14], trustworthy dissemination of data shall include the following aspects, which are then discussed with regard to the anticipated application:

- integrity of messages includes the aspect that data "is delivered as sent" and is not manipulated during the communication process,
- authenticity ensuring two aspects: The first is around the trustworthiness of the device itself, the second concentrates on the trustworthiness of the communicated data, e.g. by false injection attack.
  Basically, a mutual authentication between both sides of the communication is required. As a trade-off, one could argue that from application point of view, it is especially important to perform rapid and reliable authentication at the vehicle unit, as there, reaction times can be seen as much more critical, and abuse could lead to more serious consequences as in the case of a pedestrian.
  In order to avoid misunderstanding, the aforementioned shall not say that the situation at the other object, i.e. a pedestrian should not be verified, however, as speed and level of distraction is much lower, a higher false-alarm rate could be accepted.
- access control restricts the access to the device and to the communication elements, and is mostly a characteristics of node architecture.
- confidentiality allows only authenticated users or devices to understand the meaning of the communicated data. It is general understanding that eSafety information shall be openly shared to improve traffic efficiency and road safety [14]. Messages need to be authentic, but their contents needn't be encrypted.
- availability of the system, i.e. timely delivery is of key performance for the performance of the systems (real-time behaviour). The availability must be met under conditions of scalability (cf. ch. 3.3), but must also be secured for Denial of Service (DoS) threats. DoS may be either produced by an unintentional failure or malicious action forms a severe security risk in any distributed system.
- privacy, the protection against typical privacy-infringing malicious profiling or accidental eavesdropping.

3.2 Environmental and Architectural Conditions

Services

In VANETs two different service categories dominate: Services provisioned by a dedicated server infrastructure, in the following referred to as platform services, and decentralized ad hoc services primarily using broadcast based information dissemination, in the following referred to as VANET services [15].

Devices and Their Characteristics

The communication system consists of three types of devices: devices at the roadside (road-side unit, RSU), devices at a vehicle and devices at other objects.

It is a common assumption that devices at the roadside or the vehicle might be trusted tamper-resistant hardware security devices (HSD) that has computation capabilities to calculate private/public keys [16]. The HSD is preferably to be reconfigurable such that it can be reconfigured by down loading data from the central CA [16].

The other object might be as "simple" as a pedestrian. Simple in this context means

- it might have severe restrictions with regard to energy,
- it might have severe restrictions to cost, and
- above all, the likely situation that no connection to the internet can be taken as granted.

However, this last restriction might be relevant for all three kinds of devices. Also vehicle or road-side unit (RSU), which might be equipped with mobile communication equipment, might suffer from – at least – temporal loss of Internet connectivity.

Authorities

In many systems, a trust domain is used, which could be managed by a regional transportation authority (RTA). This trust domain allows the assignment of a system public/private key pair to each legitimate user for authentication purpose.

3.3 Functional Requirements

In addition to the objectives and characteristics described in ch. 3.1 and 3.2, two more requirements must be regarded, which are specific to CPPS

- scalability: Obviously, the number of communication nodes in the system increases significantly, if not only cars, but also infrastructure and pedestrians are communicating – and therefore must be secured.
- real-time behavior: The time conditions of the Ko-FAS systems is much more time critical than legacy C2C-communication schemes. This is valid with respect to two aspects: It concerns the time from first physical connectivity of two nodes to functional communication. But it also includes the real-time behavior during communication. For the challenges described herein, the first aspect is the more demanding.

4 Existing Solutions

4.1 Technology Scouting

A vast amount of literature can be found on the topic of network security in general [13], and of security for ad-hoc networks [17] and on the security of vehicular ad-hoc networks.

As an excellent entry point, [20] provides an extensive overview on the security of V2X networks from the Sevecom project [18]. Two aspects shall be examined in more detail, as they are crucial with regard to the requirements. Those are privacy and authentication. Other aspects, like secure routing are disregarded, as they are not relevant for the anticipated application scenarios.

4.2 Privacy Protection

The objective of privacy is the protection against typical privacy-infringing malicious profiling or accidental eavesdropping [14].

Basic Techniques

In general, location privacy protection schemes for mobile networks can be classified as policy-based and anonymity based [21].

In policy-based schemes [22], vehicles specify their location privacy preferences as policies and trust that the third party location-based service (LBS) providers adhere to these policies.

In the anonymity-based approaches [23], the location tracking of a target vehicle can be mitigated by using a randomly chosen and changing identifier, called the pseudonym [24]. Pseudonyms can either be a set of public keys, network layer addresses [25], or link layer addresses [26]. Pseudonyms are generated in a predefined way such that the adversaries cannot link a new pseudonym to previous ones of the same vehicle. The change of pseudonym denotes that the vehicle either changes its public key or addresses on the different layers (i.e., network and link). This approach regards anonymity as being untraceable between two successive locations of the target. Since pseudonyms cannot be linked to each other, they can provide a certain degree of privacy. In general, frequently changing pseudonyms are accepted as a solution to protecting the privacy of VANET [27].

In doing so, it is important to have in mind that changing the pseudonym only at one layer may still pose a risk that an attacker can link two pseudonyms from the unchanged address at the other layer [21].

To provide privacy effectively [15], the whole system architecture has to be evaluated and included in the privacy concept. In contrast to security mechanisms it is not sufficient for privacy to be included in one layer (e.g. application). To provide a high degree of anonymity the privacy concept has to ensure the so called unlinkability of "traces" a node leaves behind. Since all layers in the communication stack generate characteristic labels, such as IP-addresses, Medium Access Control (MAC)-addresses or session identifiers, the privacy component has to have control over every communication layer and the generation of labels. If any label changes all other labels have to change as well, otherwise the messages can be linked by the unchanged label(s). Hence, a single privacy component has to control all involved communication- and label-generating blocks in the system. Otherwise no reliable privacy concept can be provided.

Specific Privacy Enhancement Schemes

Various location privacy enhancement schemes are presented in [21]:

AMOEBA [50] can mitigate the unauthorized location tracking of vehicles by using the concept of group navigation for V2R communications and by introducing the random silent period between update of pseudonyms [28] for V2V communications.

In [29], the road network is divided into observed zones and unobserved zones from the viewpoint of the adversaries. Observed zones are those areas where the adversaries can track the locations of the target vehicles. The unobserved zones (also called the mix zones) are some predetermined locations (e.g., road intersections) where the vehicles vary their directions, speeds, and their pseudonyms. The adversaries would have difficulty in linking the vehicles that emerge from the mix zone to those that entered it earlier. Since the locations of mix zones are predetermined, the adversaries may still attempt to eavesdrop on transmissions originating from the mix-zone area.

In the CMIX (Cryptographic MIX-zone) [30], each vehicle obtains a public/private key pair from certificate authority (CA) via the road-side unit (RSU), and utilizes these keys to encrypt all messages while they are within the mix zone.

The concept of location k-anonymity [31] is proposed for protecting the location information through spatial and temporal cloaking. In spatial cloaking, a vehicle broadcast its coarse-grained spatial range information when the number of vehicles within its range is greater than a certain threshold. In temporal cloaking, the beacon message will not be broadcast by the vehicle until a certain number of other vehicles have visited the same location.

In [21], the concept of both grouping and mix-zone by using the neighbouring vehicle density is proposed. By monitoring the neighbouring vehicle density, each vehicle updates its pseudonym only when there are at least $k-1$ distinct neighbouring vehicles.

4.3 Authentication

Fast Authentication

In order to support the real-time behavior, authentication schemes must be fast and reliable [32]. [33] proposed a predictive key distribution approach, based on pre-distribution of the authentication key one hop ahead of the mobile station using neighbor graph knowledge, to reduce the delay caused by the authentication process. [48] proposed a fast authentication scheme using the exchange of the random nonce, which helps to generate a symmetric key for a mobile station to process authentication with a next will-be associated access point (AP). [51] proposed a pre-authentication scheme based on a hash chain, which is shared by a mobile station and an authentication server. The symmetric key used for the upcoming authentication can be pre-constructed by hashing the concatenation of an element of a hash chain from the authentication server and a handover key from the previous AP.

Pseudonymous Authentication

The basic approach to secure communication is pseudonymous authentication [35]: each vehicle (node) is equipped with multiple certified public keys that do not reveal the node identity; the vehicle uses one pseudonym at a time and after a period it abandons it and switches to a new pseudonym. This way, messages signed under different pseudonyms cannot be linked.

The pseudonyms can be obtained a priori or be reloaded, or generated by nodes on the fly [36]. In both cases, a communication transcript, under one or more pseudonyms, can be linked back to the long-term identity of the node that generated signed messages under those pseudonyms [35]. Moreover, long- or short term credentials can be revoked.

In [16], a widely used solution to handle the trade-off between authentication and privacy in VC networks is described. It is based on the digital signature of every message before transmission using a short term public/private key pair that is completely independent of both the unique identifier (IDx) of the vehicle and the long term public/private key pair (SKx, PKx) allocated to the vehicle during first registration. The long term identifier will be issued by a centralized certificate authority (CCA). The short term key pair is certified by the certificate authority (CA). The short term public/private key pair is used for signing a very limited number, normally one, of the messages. The vehicle at the end of the allowed period has to switch to another certificate. Accordingly, each vehicle is equipped with a number of certified public keys (pseudonyms) that do not reveal the vehicle identity. The long term certificate and the short term certificates are maintained by the CA. More importantly, the CA keeps a mapping between them. Since the short term keys are independent from the unique identifier of the vehicle, this mechanism guarantee that messages signed with different pseudonyms cannot be linked.

The unique identifier, the long-term keys, the short-term keys and all other cryptographic material are stored in a hardware security model (HSD) on the board of the vehicle.

Anonymous Certificates

Recently most proposals on VANET security [37] [38] [39] provide the option of using anonymous credentials in authentication while preserving traceability and revocation once such credentials are misused.

Anonymous certificates can be created by certificate issuers using blind signatures [40]. Blind signatures allow certificate issuers to digitally sign a certificate without knowing the content of the certificate [32]. A blind signature scheme can be implemented by any common digital signature scheme, such as RSA and ElGamal. To generate a blind signature, the message is firstly blinded by a user using a blinding factor (a random number). The blinded message is sent to the signer, who cannot see the real content of the message, and the signer signs it using a standard signing algorithm. The signed message is returned to the user, and then the user knowing the blinding factor can remove it from the signature. Finally, the result (message/signature) of the blind signature scheme is the same as the normal signature scheme.

Zero-Knowledge Approach

The Zero-Knowledge approach was proposed as a promising technology in [41] however, it seems that it hasn't found any acceptance since then.

5 Gap Analysis and Additional Proposals

5.1 Address Handling

The anticipated communication flow is based on individual addressing, as are many network protocols. Identification hiding based on multi- or broadcasting is not suitable for anticipated bidirectional communication between individual nodes.

Therefore, addresses for the operation in the wireless localization network shall be managed, so that identities can be disguised. However, this camouflage of addresses shall observe three important properties:

- Addresses shall be unique. It shall be avoided under all circumstances and at all times that two different nodes have one identical address.
  In order to support this, within the KoTAG-project a combination of geography-based, time-based and random addressing is proposed. Devices select addresses for a certain period of time, which include their GPS coordinates, a time-stamp and a random part.
  In order to get rid of the (already very small) probability that two devices at close locations and identical times choose the same random number, a verification of double addressing can be performed similarly to autoIP-protocol [42]. Special countermeasures can be activated if hidden-station problem impedes a direct supervision of the channel.
- Devices must be traceable during a certain period of time, so that the CPPS-functionality can be supported. Address changes can be performed at regular periods in time, however, not during a critical pre-collision phase. Intrusion detection analyses can help to discover the attempt to keep a device constantly in a critical phase so that no address change is triggered.
- Identification shall include device types, such as tag or vehicle unit. This means that one part of the identification remains unchanged throughout the lifetime of the device. However, this behavior does not endanger anonymity of the individual node.

5.2 Traceability of Certificates

One of the fundamental challenges which remain with regard to authentication is that authentication requires some kind of identification of nodes, with respect to the certification authority. Even in those cases, where devices have more than one certificate to be used, some link to the identity of the device is given, because the certification authority can store which certificate was provided to which station.

In addition, the certification authority should have some proof that the device applying for a certificate is not malicious. This requires some kind of identification or unique information of the device.

Attempts to have a multi-step approach with an additional trust center, comparable to the systems being used for credit-card authentication, help to increase the security level, however, still rely on the assumption that this additional trust center can be trusted.

5.3 Onlineliness, Revocation Lists and Scalability

Fundamentally, it is not required for the devices to maintain connectivity with the Trusted Third Party, yet it is required that the nodes in the network establish connectivity to the Trusted Third Party from time to time in order to receive certificate updates as well as information about black-listed certificates (certificate revocation list, CRL). However, the following restrictions must be taken into account, which partially are derived from ch. 3.3:

- scalability: The administration of white or black lists of certificates or devices shall be handled on an individual basis. This implies that these lists are directly dependant from the size and the lifetime of the network. Assuming a large proliferation of the system, local storage or network bandwidth and real-time behavior will be at stake. Approaches like in [43], [44] or [45] help to alleviate, but not to solve the problem.
- real-time behavior: As mentioned, real-time behavior is a key element of the anticipated functionality. Online verification of a certificate typically consumes a significant time and – of course – requires online connectivity at that very moment. This leads to a tradeoff between communication bandwidth and local storage size vs. onlineliness.
- lifetime: The more often the devices change their disguised identity the harder it will be to trace these devices. However, for each new identity, new certificates are used. This leads to another tradeoff between security and scalability. For example, the IEEE P1609.2 standard for (WAVE) proposes the issuance of short-lived certificates to reduce the overhead of CRLs. [48] proposes to distribute temporary certificates over road-side unit (RSU). Lifetime might also be restricted by number of (successful or unsuccessful) authentication attempts [49].

Of course, the attacks against mobile C2x-networks include the types of attacks that are also known for normal networks, i.e. passive and active attacks, and external and internal attacks. With regard to the requirements of ch. 3.1 and ch. 3.3, it becomes clear that an extremely hard requirement on real-time behaviour must be regarded. Therefore, at least, an intrusion detection system must be activated to detect blocking of the communication channel.

6 Outlook

Based on this technology scouting, a security architecture is currently under design and development for the KoTAG-architecture. It will be presented in the subsequent workshops of nets4cars.

Acknowledgement. This work results from the joint project Ko-TAG, which is part of the project initiative Ko-FAS, and has been partially funded by the German Bundesministerium für Wirtschaft und Technologie (Federal Department of Commerce and Technology) under contract number 19S9011. The authors are grateful for this support and for the excellent teamwork in the consortium.

References

[1] Blume, H., Flügel, S., Kunert, M., Ritter, W., Sikora, A.: Mehr Sicherheit für Fußgänger: Überblick über das vom Bundesministerium für Bildung und Forschung (BMBF) geförderte Forschungsprojekt Propedes. Elektronik Automotive (December 2011)

[2] Rasshofer, R., Schwarz, D., Gresser, K., Biebl, E.M.: Fußgängerschutz mittels kooperativer Sensorik. In: 5. Workshop FAS, Walting (2008)

[3] Andreone, L., Visintainer, F.: Prevention of road accidents involving Vulnerable Road Users: the outcomes of the WATCH-OVER European project. In: ITS 2009, Stockholm (2009)
[4] 802.11p-2010, IEEE Standard for Local and Metropolitan Area Networks - Specific requirements Part 11: Wireless LAN Medium Access Control (MAC) and Physical Layer (PHY) Specifications Amendment 6: Wireless Access in Vehicular Environments
[5] U.S. Department of Transportation, IEEE 1609 - Family of Standards for Wireless Access in Vehicular Environments (WAVE), Intelligent Transportation Systems Standards Fact Sheet (September 25, 2009), http://www.standards.its.dot.gov/fact_sheet.asp?f=80
[6] Harmonized European Standard (Telecommunication series), ETSI EN 302 571 V1.1.1: "Intelligent Transport Systems (ITS); Radiocommunications equipment operating in the 5 855 MHz to 5 925 MHz frequency band; Harmonized EN covering the essen-tial requirements of article 3.2 of the R&TTE Directive", http://pda.etsi.org/exchangefolder/en_302571v010101p.pdf
[7] Car2Car Communication Consortium, Manifesto Overview of the C2C-CC System (2007), http://www.car-to-car.org
[8] http://www.kofas.de/
[9] Lill, D., Schappacher, M., Gutjahr, A., Sikora, A.: Development of a Wireless Communication and Localization System for VRU eSafety. In: 2nd Int'l Workshop on Communication Technologies for Vehicles (Nets4Cars 2010) at 7th IEEE CSNDSP 2010, Newcastle, UK (2010)
[10] http://www.preciosa-project.org/
[11] http://www.preserve-project.eu/
[12] Lill, D., Schappacher, M., Islam, S., Sikora, A.: Wireless Protocol Design for a Cooperative Pedestrian Protection System. In: Strang, T., Festag, A., Vinel, A., Mehmood, R., Rico Garcia, C., Röckl, M. (eds.) Nets4Trains/Nets4Cars 2011. LNCS, vol. 6596, pp. 119–130. Springer, Heidelberg (2011)
[13] Eckert, C.: IT-Sicherheit. Konzepte – Verfahren – Protokolle. Oldenbourg Wissenschaftsverlag, Munich (2012)
[14] Weyl, B.: Car2Car Communication Consortium (C2C-CC): Secure Vehicular Communication: Results and Challenges Ahead. In: SeVeCom Workshop 2008, Lausanne (2008)
[15] Eichler, S.: A Security Architecture Concept for Vehicular Network Nodes. In: ICICS 2007 (2007)
[16] Peplow, R., Dawoud, D.S., Merwe, J.: Ensuring privacy in vehicular communication. In: 1st Int'l Conf. on Wireless Communication, Vehicular Technology, Information Theory and Aerospace & Electronic Systems Technology (May 2009)
[17] Davis, C.: Security Protocols for Mobile Ad Hoc Networks. VDM Verlag (2008)
[18] SEVECOM Project, Security architecture and mechanisms for V2V / V2I, Deliverable 2.1, http://www.sevecom.org
[19] Song, J.-H., Wong, V.W.S., Leung, V.C.M.: Wireless Location Privacy Protection in Vehicular Ad-Hoc Networks. In: Mobile Networks and Applications (2009)
[20] http://bbcr.uwaterloo.ca/~rxlu/sevecombib.html
[21] Song, J.-H., Wong, V.W.S., Leung, V.C.M.: Wireless Location Privacy Protection in Vehicular Ad-Hoc Networks. In: Mobile Networks and Applications (2009)
[22] Myles, G., Friday, A., Davies, N.: Preserving privacy in environments with location-based applications. IEEE Pervasive Computing 2(1) (March 2003)

[23] Gruteser, M., Grunwald, D.: Anonymous usage of location-based services through spatial and temporal cloaking. In: ACM Int'l Conf. Mobile Systems, Applications, and Services (MobiSys), San Francisco, CA (May 2003)
[24] Schoch, E., Kargl, F., Leinmüller, T., Schlott, S., Papadimitratos, P.: Impact of Pseudonym Changes on Geographic Routing in VANETs. In: Buttyán, L., Gligor, V.D., Westhoff, D. (eds.) ESAS 2006. LNCS, vol. 4357, pp. 43–57. Springer, Heidelberg (2006)
[25] Fonseca, E., Festag, A., Baldessari, R., Aguiar, R.: Support of anonymity in VANETs - Putting pseudonymity into practice. In: IEEE WCNC, Hong Kong (2007)
[26] Lei, M., Hong, X., Vrbsky, S.V.: Protecting location privacy with dynamic MAC address exchanging in wireless networks. In: IEEE Globecom, Washington, DC (2007)
[27] Garlach, M., Guttler, F.: Privacy in VANETs using changing pseudonyms - Ideal and real. In: IEEE VTC-Spring, Dublin, Ireland (April 2007)
[28] Huang, L., Matsuura, K., Yamane, H., Sezaki, K.: Enhancing wireless location privacy using silent period. In: IEEE WCNC, New Orleans, LA (March 2005)
[29] Buttyán, L., Holczer, T., Vajda, I.: On the Effectiveness of Changing Pseudonyms to Provide Location Privacy in VANETs. In: Stajano, F., Meadows, C., Capkun, S., Moore, T. (eds.) ESAS 2007. LNCS, vol. 4572, pp. 129–141. Springer, Heidelberg (2007)
[30] Freudiger, J., Raya, M., Felegyhazi, M., Papadimitratos, P., Hubaux, J.-P.: Mix-zones for location privacy in vehicular networks. In: Int'l Workshop on Wireless Networking for Intelligent Transportation Systems (WiN-ITS), Vancouver (2007)
[31] Gedik, B., Liu, L.: Protecting location privacy with personalized k anonymity: architecture and algorithm. IEEE Trans. Mobile Comput. 7(1) (2008)
[32] Zhang, C., Lu, R., Ho, P.-H., Chen, A.: A Location Privacy Preserving Authentication Scheme in Vehicular Networks. In: WCNC 2008 (2008)
[33] Mishra, A., Shin, M.H., Petroni Jr., N.L., Clancy, T.C., Arbauch, W.A.: Proactive key distribution using neighbor graphs. IEEE Wireless Communications 11(1) (2004)
[34] Wang, H., Prasad, A.R.: Fast Authentication for Inter-domain Handover. In: de Souza, J.N., Dini, P., Lorenz, P. (eds.) ICT 2004. LNCS, vol. 3124, pp. 973–982. Springer, Heidelberg (2004)
[35] Ardelean, P., Papadimitratos, P.: Secure and Privacy-Enhancing Vehicular Communication. In: Vehicular Technology Conference (2008)
[36] Calandriello, G., Papadimitratos, P., Lioy, A., Hubaux, J.-P.: Efficient and robust pseudonymous authentication in VANET. In: 4th ACM International Workshop on Vehicular Ad Hoc Networks, VANET 2007 (2007)
[37] Raya, M., Hubaux, J.-P.: Securing vehicular ad hoc networks. Journal of Computer Security, Special Issue on Security of Ad Hoc and Sensor Networks 15(1) (2007)
[38] Sun, J., Zhang, C., Fang, Y.: An id-based framework achieving privacy and non-repudiation in vehicular ad hoc networks. In: IEEE Military Communications Conference, MILCOM 2007 (2007)
[39] Lin, X., Sun, X., Ho, P.-H., Shen, X.: GSIS: A secure and privacy-preserving protocol for vehicular communications. IEEE Trans. Vehicular Tech. 56(6) (November 2007)
[40] Chaum, D.: Blind signatures for untraceable payments. In: Proc. of Advances in Cryptology (1982)
[41] Kosch, T.: Technical Concept and Prerequisites of Car-To-Car Communication. In: 5th European Congress and Exhibition on ITS, Hannover (2005)
[42] Gutjahr, A., Sikora, A.: Automatic configuration of headless, networked devices. In: Embedded World Conference, Nuremberg (2009)
[43] Papadimitratos, P., Mezzour, G., Hubaux, J.-P.: Certificate Revocation List Distribution in Vehicular Communication Systems. In: VANET 2008, San Francisco (2008)

[44] Haas, J.J., Hu, Y.-C., Laberteaux, K.P.: Design and analysis of a lightweight certificate revocation mechanism for VANET. In: 6th ACM International Workshop on VehiculAr Inter-NETworking, VANET 2009 (2009)
[45] Wasef, A., Shen, X.: EDR: Efficient Decentralized Revocation Protocol for Vehicular Ad Hoc Networks. IEEE Trans. Vehicular Technology 58(9) (2009)
[46] Kärpijoki, V.: Security in Ad Hoc Networks, http://users.tkk.fi/vkarpijo/netsec00/netsec00_manet_sec.pdf
[47] Lill, D., Schappacher, M., Islam, S., Sikora, A.: Wireless Protocol Design for a Cooperative Pedestrian Protection System. In: Strang, T., Festag, A., Vinel, A., Mehmood, R., Rico Garcia, C., Röckl, M. (eds.) Nets4Trains/Nets4Cars 2011. LNCS, vol. 6596, pp. 119–130. Springer, Heidelberg (2011)
[48] Wang, J., Jiang, N.: A simple and efficient security scheme for vehicular ad hoc networks. In: IEEE Int'l Conference on Network Infrastructure and Digital Content, IC-NIDC 2009 (2009)
[49] Sun, J., Fang, Y.: A defense technique against misbehavior in VANETs based on threshold authentication. In: IEEE Military Communications Conference, MILCOM 2008 (2008)
[50] Sampigethaya, K., Li, M., Huang, L., Poovendran, R.: AMOEBA: robust location privacy scheme for VANET. IEEE J. Select. Areas Commun. 25(8) (2007)
[51] Hong, K., Jung, S., Felix Wu, S.: A Hash-Chain Based Authentication Scheme for Fast Handover in Wireless Network. In: Song, J.-S., Kwon, T., Yung, M. (eds.) WISA 2005. LNCS, vol. 3786, pp. 96–107. Springer, Heidelberg (2006)

ICDMS: An Intelligent Cloud Based Disaster Management System for Vehicular Networks

Zubaida Alazawi[1], Mohmmad B. Abdljabar[2], Saleh Altowaijri[1], Anna Maria Vegni[3], and Rashid Mehmood[4]

[1] School of Engineering Swansea, Wales, UK
e_zubaida97@yahoo.com, salehmsm@hotmail.com
[2] Al-Mustansereya University Baghdad, Iraq
mbakr67@yahoo.com
[3] University of Roma Tre Rome, Italy
amvegni@uniroma3.it
r.mehmood@gmail.com

Abstract. The importance of emergency response systems cannot be overemphasized today due to the many manmade and natural disasters in the recent years such as September 2001 and the recent Japan earthquake and tsunami disaster. The overall cost of the Japan disaster alone is estimated to have exceeded 300 billion USD. Transportation and telecommunications play a critical role in disaster response and management in order to minimize loss of human life, economic cost and disruptions. Our research is concerned with developing emergency response systems for disasters of various scales with a focus on transportation systems, which exploit ICT developments.

In this paper, we leverage Intelligent Transportation Systems including Vehicular Ad hoc Networks, mobile and Cloud Computing technologies to propose an intelligent disaster management system. The system is intelligent because it is able to gather information from multiple sources and locations, including from the point of incident, and is able to let vehicles make effective strategies and decisions of communication protocols usage. Hybrid vehicular communications based on vehicle-to-vehicle and vehicle-to-infrastructure protocols are opportunistically exploited. The effectiveness of our system is demonstrated through modelling the impact of a disaster on a real city transport environment and comparing it with the case where our disaster management system was in place. We report great benefits derived from the adoption of our proposed system in terms of improved and balanced traffic flow and smooth evacuation.

Keywords: Disaster Management and Resilience, Intelligent Transportation Systems, Cloud Computing, VANETs, Mobile Communications.

1 Introduction

The importance and scope of emergency response systems have grown tremendously over the past decade in particular after September 11, 2001. Disasters, manmade and natural, are a cause of great economic and human losses each year throughout the world. The overall cost of the recent Japan earthquake and tsunami disaster alone is estimated to have exceeded 300 billion USD. This has driven many new initiatives

A. Vinel et al. (Eds.): Nets4Cars/Nets4Trains 2012, LNCS 7266, pp. 40–56, 2012.

and programs in countries throughout the world, in particular in the US, Europe, Japan and China [1], [2], [3].

A great deal of advisory and policy documents has been developed by various government authorities through surveys, consultations, experiences and other means of research [4], [5]. As an instance, ensuring the success of mass evacuations is a critical aspect for the Department of Transportation, in cooperation with the Department of Homeland Security. The main purpose is the assessment of mass evacuation plans for the country's most high-threat, high-density areas, as well as the identification and prioritization of deficiencies on those routes that could impede evacuations [6].

Transportation and ICT (Information and Communication Technologies) technologies play critical roles in responding to emergencies and minimising disruptions, human and socioeconomic costs. We have witnessed unprecedented advancements in information and communication technologies over the last few decades and the role of ICT technologies in Intelligent Transportation Systems (ITS) is to grow tremendously. Vehicular Ad hoc Networks (VANETs), sensor networks, social networks, Vehicle-to-Vehicle (V2V) and Vehicle-to-Infrastructure (V2I) technologies are enabling transformational capabilities for transportation [7], [8]. The ability to monitor and manage transportation system in real-time and at high granularity has grown tremendously due to sensor and vehicular network that generate huge amount of extremely useful data. However, many challenges in realising the ITS potential remain, including the interworking and integration of multiple systems and data to develop and communicate a coherent holistic picture of transportation systems. This is particularly difficult given the lack of data and systems interoperability as well as the business models to develop such an advanced infrastructure, which requires coordination between many stakeholders and general public.

Cloud Computing has emerged as a technology, coupled with its innovative business models, which has the potential to revolutionise the ICT and ITS landscape. It is already making a huge impact in all sectors through its low cost of entry and high interoperability [9]. Moreover, the technology allows one to develop reliable, resilient, agile, and incrementally deployable and scalable systems with low boot-up time, and at low costs, while giving users access to large shared resources on demand, unimaginable otherwise.

In this paper, we leverage the advancements in the ICT technologies —including ITS, VANETs, social networks, mobile and Cloud Computing technologies— to propose an Intelligent Cloud based Disaster Management System (ICDMS) for an urban environment. By exploiting these latest technologies, the system is able to gather information from multiple sources and locations, including from the point of incident, and is able to make effective strategies and decisions on communication protocols in order to propagate the information to vehicles and other nodes in real-time.

This work aims to (i) improve the system model by means of introducing a novel message propagation algorithm, and (ii) validate the effectiveness of our proposed ICDMS, by means of extended simulation results. The previous work had only reported the road traffic modelling results for disaster management [10]. The effectiveness of our system is demonstrated through modelling the impact of a disaster on a real city transport environment [1]. We model two urban scenarios for disaster management (i) using

[1] This paper focuses on disaster management although our work is broadly applicable to emergency response systems of any scales, such as emergency vehicles warning system, and, in general to the operational and strategic management of transportation systems.

traditional technologies, and (ii) exploiting our intelligent, modelling and vehicular Cloud-based disaster management system, ICDMS. The comparison of the two scenarios demonstrates the effectiveness of our system in terms of the number of people evacuated from the city, the improved traffic flow and a balanced use of transportation resources.

This paper is organised as follows. Section 2 provides introduction to Cloud Computing and VANETs. Section 3 describes the architecture and various components of the proposed ICDMS. Section 4 provides an evaluation of our system and Section 5 concludes the paper with directions for future work.

2 Background Material

2.1 Cloud Computing

Cloud Computing (CC) is swiftly becoming a very attractive and foundational element in global enterprise computing [9]. It is believed that the foundation of CC was a result of a set of many pre-existing and researched concepts, i.e. distributed and grid computing, service oriented architectures, virtualisation, and Software as a Service [13], [14]. CC is an enabler of innovation and new business models in enterprise computing. There are several companies across a wide range of industries which implement, develop and offer cloud technologies. CC can be defined as web applications and server services that users pay for in order to access, rather than software or hardware that the users buy and install themselves [11]. CC offers great advantages for organisations and businesses of all sizes. One of the main advantages of is the fact that organisations and businesses do not need the infrastructure or the necessary information and knowledge to develop or maintain the infrastructure, as providers are taking care of all that. Moreover, CC also allows users to access a wide variety of applications, services, and hardware, which they might not be able to access otherwise.

By using CC, organisations and businesses hugely save money and can cut the cost because they will rent the software and applications rather than buying them and have them installed on their machines. CC could also save time due to the fact that businesses will not have to install and or upgrade software and applications, as well as businesses can easily gain access to applications which are specified to their needs and available only over the Internet [11], [12].

The true innovation and improvement that Cloud Computing brings in is seen in the way its computing services are being provided to the customers. Such services comprise the ability to offer software applications, programming platforms, data storage or computing infrastructure. There are several IT system components that are being offered as services, such as Hardware as a Service (HaaS), Software as a Service (SaaS) and Data as a Service (DaaS). These services are offered together as a Platform as a Service (PaaS) [13], [14]. It is therefore useful to look at Cloud Computing as a stack of layers, as depicted in **Fig. 1**, [14]. Notice that the cloud computing systems comprise five layers, i.e. applications, software environments, software infrastructure, software kernel, and hardware.

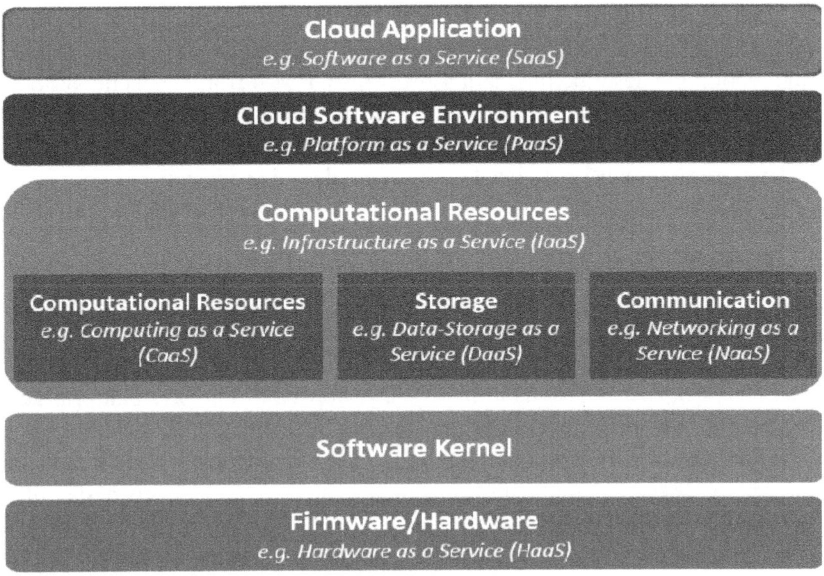

Fig. 1. The layers of Cloud Computing

2.2 VANETs and Intelligent Transportation Systems

The inherent human desire for change, progress, mobility, entertainment, safety and security are leading the way to the development of Intelligent Transportation Systems (ITS). Vehicular Ad-hoc NETworks (VANETs) are the most prominent enabling technology for ITS. VANETs are formed on the fly by vehicles equipped with wireless communication capability. The participant nodes in VANETs (i.e. vehicles) interact and cooperate with each other by short-range direct communications, by hoping messages through vehicles (Vehicle-to-Vehicle) and road side masts (Vehicle-to-Infrastructure). Traditionally, information about traffic on a road is only available through inductive loops, cameras, roadside sensors and surveys. VANETs provide new venues for collecting real-time information from onboard sensors on vehicles and for quick dissemination of information. The information collected through individual vehicles participating in VANETs can be integrated together to form a real time picture of the road situation. Many new applications have been enabled through VANETs, though safety and transportation efficiency applications are the most important driver for VANETs. The various ITS stakeholders such as governments, telecommunication companies and car manufacturers are working together to make VANETs based ITS a reality. Hundreds of projects are underway in the US, Europe, Japan, China, Singapore and other countries in the world helping with research, innovation, testing and standardisation activities [15], [16].

3 The Intelligent Cloud-Based Disaster Management System

In this section we describe the main aspects of ICDMS, from a general system architecture overview in Subsection 3.1, up to a detailed analysis of various components in Subsection 3.2. In particular, we address on the Intelligent Layer, which describes how data information is propagated in the vehicular environment.

3.1 System Architecture

Fig. 2 depicts the system architecture of the Intelligent Cloud-based Disaster Management System. The system consists of three main layers:

1. The *Cloud Infrastructure as a Service* layer provides the base platform and environment for the intelligent emergency response system;
2. The *Intelligence Layer* provides the necessary computational models and algorithm in order to devise optimum emergency response strategies by processing of the data available through various sources;
3. The *System Interface* acquires data from various gateways including the Internet, transport infrastructure such as roadside masts, mobile smart phones, social networks etc.

As depicted in **Fig. 2**, vehicles interact with the gateways through Vehicle-to-Vehicle (V2V) or Vehicle -to-Infrastructure (V2I) communications. For example, vehicles may communicate directly with a gateway through Internet if the Internet access is available. A vehicle may communicate with other vehicles, road masts, or other transport infrastructure through point-to-point, broadcast or multi-hop communications.

The emergency response system provides multiple portals or interfaces for users to communicate with the system, i.e. the Public Interface, the Transport Authorities Interface, and the Administrators' Interface. The Public Interface allows any individual to interact with the system. The purpose is to interact with the system on one-to-one or group/organisation basis with the system, either to request or provide some information. Of course, an authentication, authorization and accounting system is expected to be in place to allow and control various activities and functions. The Transport Authorities Interface is a high-privilege interface for the transport authorities to affectively manipulate the system for day-to-day operational management. The Administrators' Interface provides the highest privilege among the system users and is designed for policy makers and strategists to enable highest level system configuration.

The motivation and background for a Cloud-based distributed control system can be found in our earlier work [17], where we presented an architecture for distributed virtualisation using the Xen hypervisor. It allows control and management of a distributed system by posting high-level queries on the system and their validation through real-time monitoring and control of the system.

Monitoring relates to the acquisition of data and control relates to despatch of commands and decisions. Moreover in [18], Rao and Diwanji propose a Pervasive Cloud using the WiMAX broadband technology for railway infrastructure.

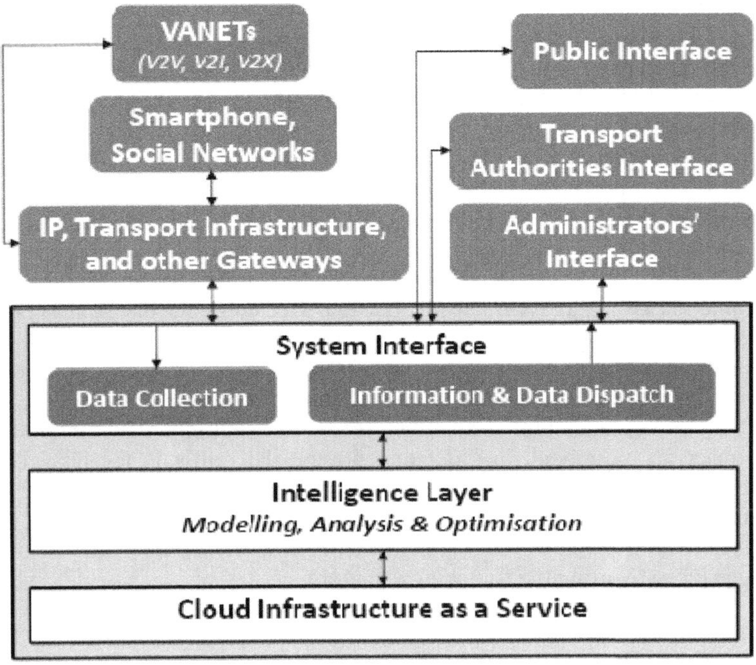

Fig. 2. Intelligent Cloud-based Disaster Management System – Proposed Architecture

3.2 The Intelligence Layer

As mentioned in Subsection 3.1, our system has a System Interface that communicates with various user interfaces and communication gateways. This interface is used to gather and propagate data, information and decisions in order to carry out day-to-day transport management operations, policy implementations, and emergency response operations. The Intelligence Layer consists of various mathematical models, algorithms and simulations, both stochastic and deterministic. These models accept transport related data received from various sensors such as inductive loops, intra-vehicular sensor networks, V2V/V2I communications, and user interfaces. The data received from various sources goes through an internal validation layer before it is accepted by the modelling and analysis layer. The modelling or simulation algorithm is used for a particular activity based on the nature of the activity. In some cases, it is necessary and/or affordable to employ microscopic traffic models, for example in developing transport policies and procedures; this is due to the demands on higher accuracy and greater flexibility on the available time for decision, optimisation and analysis. In other cases, microscopic simulations may not be necessary, or may not be possible, due to the real-time nature of operations such as day-to-day transport management operations.

Emergency response systems are an extreme example because, (i) the availability of real-time data may be greatly limited due to the unavailability of many communication sources due to a disaster (e.g. broken communication links), and (ii) the time period in which the system has to act would be short. In such cases, macroscopic models, which require relatively small computational time and resources, may be the only option. Different types of modelling are assumed for particular purposes, like the models in [15], [20], [21] for vehicular grid networks, and in [22] for Markov modelling of large systems, and finally in [23] for 3D virtual reality microscopic simulator. However, which models to invoke in a particular situation is an area of ongoing investigation and we will continue to improve on our automatic model selection algorithm. We will also be looking at ways of enhancing our distributed algorithms so we could invoke the most precise models for real-time critical situation such as great disasters. It is important to note here that we envisage a Cloud infrastructure that is virtualised and flexible to exist, or moved, outside the area affected by the disaster. This is possible considering the capability of Cloud technology.

In this paper we use a macroscopic model that is particular suitable for emergency situations, where time to act is short and real time information is limited due to possibly broken communication links. We consider the Lighthill-Whitham-Richards (LWR) macroscopic model, in order to represent the traffic in the city. The LWR model can be used to analyse the behaviour of traffic in road sections, and describe the dynamic traffic characteristics such as speed (u), density (ρ), and flow (q)[2]. The model is derived from the conservation law (first order hyperbolic scalar partial differential equation) by using the following equation:

$$\frac{\partial \rho}{\partial t} + \frac{\partial \rho u}{\partial x} = 0 \tag{1}$$

where ρ is the traffic density in vehicle/km, and u is the traffic velocity according to distance x and time t. By using Greenshield traffic model, the relation between ρ and u could be as follows

$$u\,(\rho) = u_{max} = \left(1 - \frac{\rho}{\rho_{max}}\right) \tag{2}$$

where umax is the maximum speed, and ρmax is the maximum density. The fundamental relationship between flow, density, and speed is given by

$$q = \rho \cdot u. \tag{3}$$

Message Propagation via Hybrid Communications

In our system, the Intelligent Layer also addresses on how to forward messages in the vehicular environment in an optimal way[3]. For this aim, we consider a hybrid vehicular communications paradigm which results to be very suitable for achieving seamless connectivity in VANETs, wherein changes in the dynamic topology, vehicles' speed, as well as traffic density affect vehicle communications.

Connectivity management represents a challenge for VANETs, and more in general in disaster scenarios, where the network infrastructure is assumed to be not available in specific areas, as well as vehicle density can quickly change. Exploiting both V2V and V2I represents an effective integrated solution for avoiding disconnections and guaranteeing seamless data communication [24], [25]. In [24] the proposed

---

[2] In this paper, we will use flow and volume interchangeably.

[3] By means of opportunistically connecting vehicles whenever links are available.

approach, i.e. Vehicle-to-X (V2X), provides efficient message delivery in a vehicular grid, where vehicles opportunistically select the most viable communication protocol among V2V and V2I, and can switch from V2V to V2I, and vice versa.

In this paper we adopt the broadcast V2X approach from [24] into the ICDMS structure for emergency scenarios. The cooperation and coexistence of V2V and V2I can assure a good connectivity such scenarios, especially in sparsely connected neighbourhoods where V2V communications are not always feasible. V2X can reduce the time required by a message to propagate from a source vehicle to the farthest vehicle inside a certain strip-shaped area-of-interest. Moreover, it represents a smart and realistic communication protocol, reduces the message delivery time, as well as avoids disconnections due to changed traffic density and dynamic topological changes. V2X is a pragmatic broadcast protocol where vehicles' actual transmission data rates are subjected to continuous changes due to physical obstacles, vehicle density, speed, network overload, etc.

Vehicles communicating via V2X get experience of local information, assumed as global. Each vehicle continuously monitors its local connectivity by storing HELLO broadcast messages. A vehicle will be aware of neighbouring wireless networks on the basis of broadcast signalling messages sent by the Road Side Units (RSUs), as well as will be able to determine if it is within a cluster or is travelling alone on the road.

Let us assume, for each connectivity link from the i-th to the j-th vehicle, the link utilization time τ(i;j) [s] as the time needed to transmit a message of length L [bit] from the i-th to the j-th vehicle, at an actual data rate f(i;j) [Mbit/s], such as[4] .:

$$\tau_{(i,j)} = L/f_{(i,j)} \tag{4}$$

Let us define a path from i-th source vehicle to k-th destination node (i.e. vehicle or RSU), comprising of a sequence of M hops. Each hop can be a link between two neighbouring vehicles via V2V or from a vehicle to a RSU via V2I. From the definition of a path, we define the path utilization time T(i, k) [s] from the i-th source vehicle to the k-th destination node as the sum of single link utilization time parameters for each hop that constitutes the path as:

$$T_{(i,k)} = \tau_{(i,j)} + \tau_{(i,x)} + \cdots + \tau_{(x,k)} == L \sum \left[f_{(i,x)}^{-1} \right]. \tag{5}$$

Among all the paths from a source vehicle to a selected destination node, the optimal path will be the one with the minimized path utilization time, such as

$$\min T_{(i,k)} = L \cdot \min \sum \left[f_{(i,x)}^{-1} \right]. \tag{6}$$

Based on the estimation of the path utilization time (i.e., the message delivery time for a path from a source vehicle to a destination node), V2X is used to reduce the amount of hops needed to deliver the message. In disaster scenarios, where time is limited and connectivity links are possibly broken, a fast communication protocol which exploits time limited hops represents a viable approach.

[4] Notice that such definition applies also for connectivity links from a vehicle to a Road Side Unit.

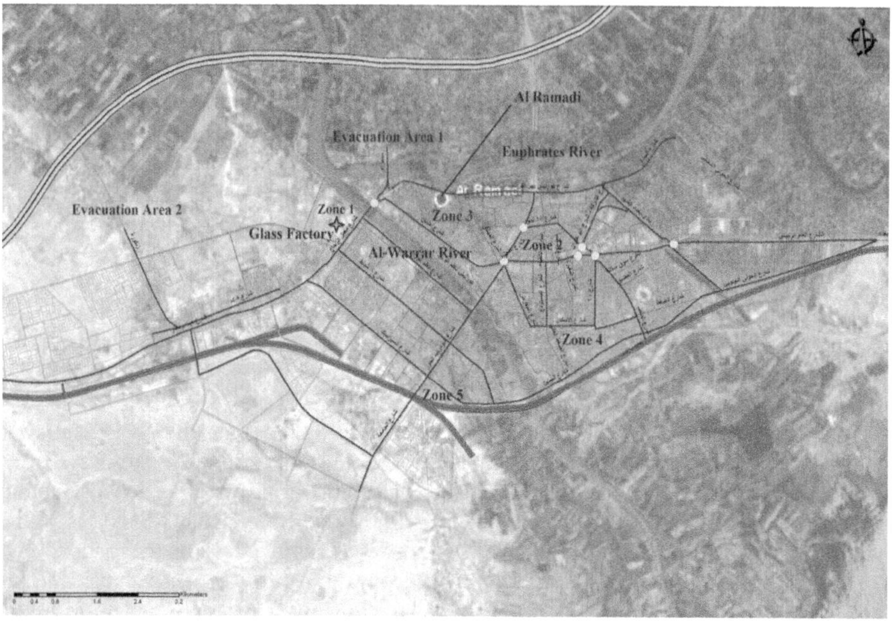

Fig. 3. Transportation network of the Ramadi city

4 System Evaluation

After described the mathematical description of the macroscopic model and the hybrid communication protocol V2X, in this section, we will describe our approach for city evacuation in a disaster emergency situation.

The transportation network of the Ramadi city consisting of zones, nodes, and links is depicted in **Fig. 3**. We now make use of our earlier discussions in Section 3 and describe the disaster scenario in Section 4.1 and 4.2. Subsequently, in Section 4.3, we present analysis of the system and establish its usefulness as a disaster response system.

4.1 The Ramadi City and Its Transportation Network

The Ramadi city (Al Ramadi) is the capital of Al Anbar Governorate and is situated at the intersection of the Euphrates River and Al Warrar Channel. The Habbaniya Lake is located a short distance to the south of the City of Al Ramadi. The General Directorate of Physical Planning of the Ministry of Municipalities and Public Works (MMPW) is preparing for the Development Strategy of Al Ramadi, and it is in line with the development policies of MMPW for other Iraqi cities. The Association of the Canada-based HYDROsult Center for Engineering Planning (HCEP) and the Iraq-based Engineering Consultancy Bureau of Al Mustansiriya University has been commissioned by MMPW to carry out the tasks of this assignment and they have produced a second stage report [16] in November 2009 for the development of the

Ramadi city. Iraq is now open to new developments and it is a great opportunity to develop intelligent transportation systems for Ramadi and other Iraqi cities.

Fig. 3 shows the transportation network map of the Ramadi city, the network consists of zones, nodes, and links. The city is divided into 5 traffic zones; Zone 1 and Zone 5 are in the west side of Al-Warrar River that divides the city into two parts. Zone 1 represents the location of a huge glass factory; Zone 5 represents the west part of the city. The east part of the city contains Zones 2, 3, and 4. Zone 2 represents the old city centre that attracts high number of trips in the morning peak hour.

Note also in **Fig. 3** the two evacuation areas, Evacuation Area 1 and Evacuation Area 2. Their purpose is to provide an appropriate and safe location for the population in case a major disaster strikes the city and people need to be moved out of the city. The two evacuation zones are chosen in the north of the city, because there is an international roadway that joins the Iraqi borders with Syria and Jordan in the west with the capital Baghdad, and the evacuation zones are just a few minutes away from that road. In the south of the city there is the desert only and the connections of roadways in this area are very poor. If there is a need to give medical supplies or transport to injured and affected population, it will be best provided through the international roadway. The area in the east side of the city is mostly is for agriculture land use and is a private property. The west street leads to a nearby city about 30 km away, this city has a good hospital but it can be best reached through the northern international roadway. We will discuss the city network and evacuation plan further in Section 4.

Table 1. An O-D Matrix of the transportation network in Ramadi city

	zone 1	**zone 2**	**zone 3**	**zone 4**	**zone 5**
zone 1	0	0	0	0	0
zone 2	82	0	172	935	228
zone 3	172	2757	0	1171	108
zone 4	343	10026	381	0	248
zone 5	272	4835	358	1699	0

4.2 The Disaster Scenario

Let us consider **Fig. 3**, which divides the Ramadi city into 5 zones. In Table I, we quantify transportation trends of the city in terms of an Origin-Destination (O-D) matrix between the five city zones. The numbers of trips in the O-D matrix shown are in the mid-week period with natural conditions. These trips are calculated using the Fratar model. Note that the highest rate of trips is toward destination Zone 2 in the city centre.

In Zone 1 lies a glass factory and beside it is Al-Warrar dam; both of these pose major risks to the city. We consider in this paper the risks related to the glass factory and Al-Warrar dam as a case study in order to describe and evaluate our emergency response system. The related potential risks for disaster events in Zone 1 are outlined below:

- Fire hazards at the main factory;
- Technology failure due to shutdown of power plants that feed the city;
- Explosion of hazardous materials in the glass factory;
- Terrorist attack in the area of the factory;
- The collapse of Al-Warrar dam adjacent to the factory.

The above listed disaster events except the last one may last several hours before it will be controlled; for transportation planning purposes, special care is required to handle the emergency situation and saving peoples' lives.

We now focus on a disaster event, which could happen in the glass factory. This event could be any one of the potential risks listed earlier in this section, e.g. fire or explosion of hazardous materials in the glass factory. The details of the event are as follows.

The traffic conditions in a city typically vary during the course of the week. We consider that the event happens during the mid-week period, e.g. on Tuesday. Our methodology is independent of a particular day/time, although the traffic situation would vary depending on the day and time of the disaster event. Usually the most critical condition in the traffic network is in the morning peak (herein between 7:30 am to 8:30 am) and evening peak hour (2:00 pm to 3:00 pm). These peaks are for official commuters but the commercial activity in the city centre usually begins after 9:00 am, at this time the peak hour are somehow relieved.

We consider that the incident happens at 9:30 am. The event causes the network to be closed in the Zone 1 and some nearby road links. An emergency response system is required, at this stage, to coordinate the city transport, communicate with the city population, and lead people out of the city to a safe location. In this case of the Ramadi city, people will be moved to the two evacuation areas.

4.3 Results and Discussion

We consider and compare two scenarios for emergency response system, such as:

1. The *traditional disaster response system* where people will gain awareness of the disaster situation and response procedure through media such radio, television, telephones (given that such means are still accessible), and through their physical environment (*e.g.* interacting with the people who are in the nearby area);
2. Our proposed ICDMS, which automatically collects data; intelligently processes the data; and, devises and propagates effective strategies and decisions based on the real-time situation, in line with appropriate policies and procedures already in place in the system. Messages are forwarded according to the V2X hybrid vehicular approach [24].

We evaluate the two systems and compare their performance in the following subsections.

A Traditional Disaster Response Scenario

A disaster usually causes most people who are in its vicinity to move away from the disaster location. The panic sets off and people start pushing each other without any effective coordination. The situation with vehicles becomes no different, as in the absence of any effective coordination, the roadway sections around a disaster area are blocked, and the incident will spill over like a shockwave over the entire network system. Such a reaction for the Ramadi city caused by a major disaster in the glass factory is depicted in **Fig. 4**. As we can notice, the roads, near the factory, that connect Zone 1 with Zones 3 and 5, and with Evacuation Area 1, all are blocked (depicted by the roads coloured in black). Also note that the roads connecting Zones 2, 3 and 4 with each other, all have very low volume below 500 vehicles per hour (depicted by the roads coloured in red). We further note a couple of roads near Zone 2, nearer the outer boundaries of the city, with volumes between 500 and 1000 vehicles per hour (represented by roads coloured in blue). Furthermore, the roads located at the outer boundaries of the Ramadi city are coloured in brown and green, depicting higher volumes, between 1000 and 1500 (brown), and greater than 1500 vehicles per hour (green), respectively.

The vehicle volumes that we have computed using our models amounts to 660 vehicles per hour (400 vehicles in the first 30 minutes) after the glass factory incident for Evacuation Area 1, and 2200 vehicle per hour (1000 vehicle in the first 30 minutes) for Evacuation Area 2. Clearly, there are many more vehicles (almost 4 times) reaching Evacuation Area 2 compared to Evacuation Area 1.

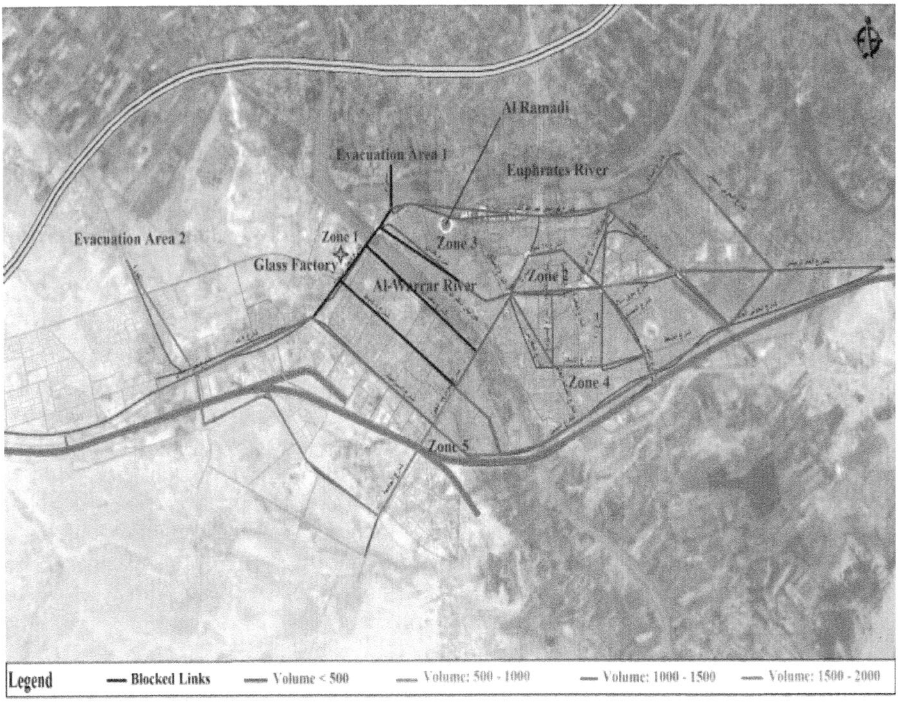

Fig. 4. Transportation network of the Ramadi city with traditional disaster response system

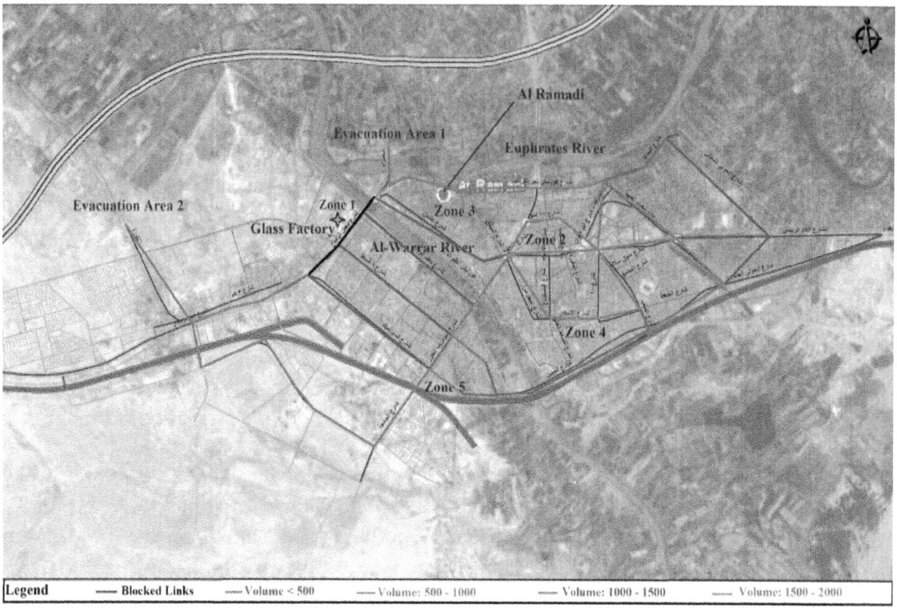

Fig. 5. Transportation network of the Ramadi city with our proposed disaster management system

The traffic situation in the city network as depicted in **Fig. 3** is calculated using the macroscopic model. It represents a snapshot taken at 10 a.m, i.e. half an hour after the disaster incident has taken place. The 30 minutes period after the incident gives some opportunity to people to start heading towards, and reaching safe places (such as evacuation areas) outside the boundaries of the city. This period also gives time for transmission of the information so most of the road users know what is happening and where they should be heading.

A final note on the evacuation process: the public transportation vehicles in Ramadi city consist of buses only. The public transportation vehicles will be involved, where possible, in the cases of emergencies, although it will need a decision from the City Council authorities. There is however a plan in place for emergency events in the Ramadi city that 50% of the public vehicles will be involved in transporting people to safe areas. Since public transportation vehicles have a high passenger capacity, these will be useful in the evacuation process. Thirty minutes after the event, i.e. when all the vehicles in the event area have received the warning information messages, is an appropriate period to make such a decision of incorporating these public vehicles in the evacuation process. Notice that in the traditional disaster response scenario, vehicles are communicating each other via V2V, and to the network infrastructure via V2I, whenever connectivity links are available.

Intelligent Cloud Based Disaster Management System

We now evaluate our proposed disaster response system. All the disaster scenario conditions are same as in the previous section including the role of public transportation in the evacuation process. The difference lies in the ability of the system to (i) acquire real-time data, and establish communication through V2X hybrid protocol, smartphones and social networks, (ii) process the data and devise an optimum strategy by data analysis, and (iii) coordinate and control road traffic and other efforts through dissemination of information and management of the available transport infrastructure (e.g. controlling traffic signals if possible, sending a route map to the traffic navigators and other GPS enabled devices etc). These three steps are iterative and can provide a periodic update to take any real-time changes into account. For instance, the macroscopic modelling in the Intelligence Layer is used to periodically compute an O-D matrix depending on the type of disaster and real-time traffic conditions. The O-D matrix then can form the basis of information that is propagated to the transport infrastructure and authorities, individuals and groups. Furthermore, certain boundary conditions will be enforced on the city through traffic management systems and authorities, such as (i) there will be no entry in the city, (ii) no entry in the area of event, (iii) many routes will be changed into one way flow outside the event area etc.

The road traffic network situation after the disaster hits the Ramadi city is depicted in **Fig. 5**, this time though we have exploited our proposed disaster management system to curtail the disaster impact. As in **Fig. 4** the network represents a snapshot taken at 10 am, half an hour after the disaster incident has taken place. Comparing **Fig. 4** with **Fig. 5**, we note less black and red, and more green roads in **Fig. 5**. For the ICDMS the roads leading to both evacuation areas are green representing clear roads and high flow (1500-2000 vehicles per hour). Moreover, we can notice a greater part of the city centre has roads with free flow (i.e. in green colour) except the roads between Zone 1 and Zone 5 which are coloured in red, representing low flow at less than 500 vehicles per hour. The road next to the glass factory is coloured black and represents a broken link. Also, a few roads near Zone 2, nearer the outer boundaries of the city, with low (red), medium (blue) and medium high (brown) volumes. The low vehicle volume (less than 500) is because of the use of alternative roads available in this case towards Evacuation Area 1.

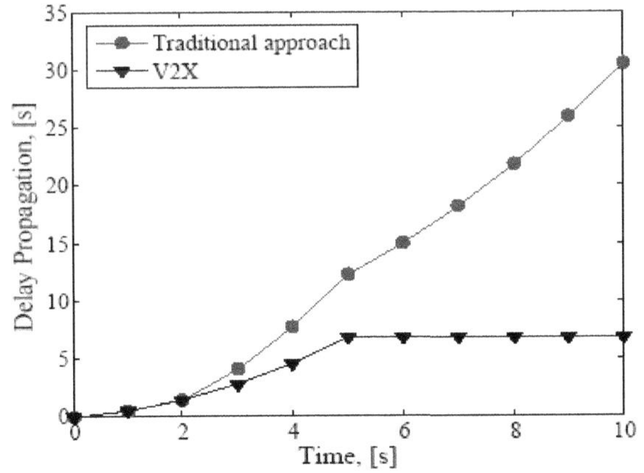

Fig. 6. Delay propagation in the transportation network of the Ramadi city. Comparison between V2X used in ICDMS and the traditional approach for disaster management.

Based on the computations and our models, 2660 vehicles per hour (1260 vehicles in the first 30 minutes) are being evacuated to Evacuation Area 1, and 2860 vehicle per hour (1530 vehicle in the first 30 minutes) are evacuated to Evacuation Area 2. The evacuation vehicle volume per hour is almost similar for both evacuation areas. This is clearly a balanced use of the two evacuation areas, an improvement over the traditional disaster management approach reported earlier where the use of Evacuation Area 1 was significantly smaller.

Finally, in **Fig. 6** we present simulation results for the average delay propagation [s] that occurs with V2X, adopted in ICDMS, and the traditional multi-hop technique for disaster manager. It verifies the effectiveness of our approach –in terms of a reduction of delay– as compared with traditional opportunistic networking scheme in VANET.

We simulated the propagation of warning messages from a vehicle near the glass factory, where the incident happens at 9:30 am. In particular, for ICDMS approach, the following events occur: (i) just after the incident has occurred, a source vehicle travelling in the S direction from the glass factory sends a message along on the same direction; (ii) at t = 4 s the message is propagated multi-hop within a cluster in the S direction; (iii) at t = 7 s a relay vehicle enters an RSU's radio coverage, and the message is transmitted via V2I to the RSU, until it will be received by other vehicles at t = 10 s.

We compared this scenario with traditional opportunistic networking technique in VANETs, where the following events occur: (i) just after the incident has occurred the source vehicle travelling in the S direction from the glass factory sends a message along on the same direction; (ii) at t = 4 s the message is forwarded to a vehicle in the N (opposite) direction; (iii) at t = 6 s the message propagates via multi-hop within a cluster in the S direction; (iv) the transmission stops at t = 10 s.

For comparative purposes, in the simulation setup we posed parameters according to [24]. We assume the source vehicle is moving at 70 km/h, the inter-RSU and inter-vehicular distance is 500 and 100 m, respectively. Typical message size L = 300 bit, and data rate transmission B = 10 Mbit/s (e.g., for WiMax connectivity), have been assumed. The inter-vehicle communication data rate has been assumed equal to 6 Mbit/s, while data rates via V2I and I2V are in the range [2.5, 14] Mbit/s.

We notice a strong reduction in the delay propagation with respect to other forms of opportunistic networking: after t = 10 s from the incident occurrence, the maximum transmission delay in ICDMS system is 6.81 s, while 30.45 s for the traditional multi-hop approach. The reduction of delay in ICDMS is mainly due to the protocol switching decision of V2X, which exploits high data rates from wireless network infrastructure.

5 Conclusions and Future Work

The importance of emergency response systems cannot be overemphasized due to the many manmade and natural disasters in the recent years. A greater penetration of ICT in ITS will play a critical role in disaster response and transportation management in order to minimize loss of human life, economic costs and disruptions. In this paper, we exploited ITS, hybrid V2V/V2I protocols, mobile and Cloud Computing technologies and proposed an Intelligent Cloud-based Disaster Management System (ICDMS). The system architecture and components were described. ICDMS was evaluated using modelling and simulations and its effectiveness was demonstrated in terms of improved disaster evacuation characteristics. Future work will focus on further analysis and validation of the disaster management system, and on broadening the scope of this work to real-time operational and strategic management of transport infrastructure.

References

1. Vásconez, K.C., Kehrli, M.: Highway Evacuations in Selected Metropolitan Areas: Assessment of Impediments. Techical Report: FHWA-HOP-10-059, Federal Highway Administration, Office of Transportation Operations, Washington, DC 20590, p. 107 (2010)
2. Owens, N., et al.: Traffic Incident Management Handbook. Science Applications International Corporation (SAIC): 8301 Greensboro Drive McLean, VA 22102, pp. 116 (2010)
3. RITA | ITS | Emergency Transportation Operations (June 27, 2011), `http://www.its.dot.gov/eto/eto_approach.html` (accessed: June 27, 2011)
4. Metropolitan Government of Nashville and Davidson County, Tennessee, Emergency Preparedness Survey for Davidson County (2008)
5. Drake, R.: The Hierarchy of Emergency Preparedness. In: Hakim, S., Blackstone, E.A. (eds.) Safeguarding Homeland Security, pp. 31–40. Springer, New York (2009)
6. Departments of Transportation & Housing & Urban Development and Related Agencies Appropriations Act, 2010 Conference Report (111-366) to Accompany HR 3288 & Public Law 111-117, FY 2010 Consolidated Appropriations Act (2010)
7. Buchenscheit, A., Schaub, F., Kargl, F., Weber, M.: A VANET-based emergency vehicle warning system. 2009 IEEE Presented at the Vehicular Networking Conference (VNC), pp. 1–8 (2009)
8. Rizvi, S.R.A., Olariu, S., Rizvi, M.E., Weigle, M.C.: A Traffic Chaos Reduction Approach for Emergency Scenarios. In: IEEE International Performance, Computing, and Communications Conference, IPCCC 2007, pp. 576–578 (2007)
9. Furht, B., Escalante, A. (eds.): Handbook of Cloud Computing, 1st edn. Springer
10. Alazawi, Z., Altowaijri, S., Mehmood, R., Abdljabar, M.B.: Intelligent disaster management system based on cloud-enabled vehicular networks. In: Proc. of 11th International Conference on ITS Telecommunications (ITST), August 23-25, pp. 361–368 (2011)
11. Chee, B.J.S., Curtis Franklin, J.: Cloud Computing: Technologies and Strategies of the Ubiquitous Data Center. CRC Press, Inc. (2010)
12. Murray: Enterprise grade cloud computing. In: Proceedings of the Third Workshop on Dependable Distributed Data Management, Nuremberg, Germany, p. 1 (2009)
13. Leimeister, S., Riedl, C., Krcmar, H., Böhm, M.: Cloud Computing - Outsourcing 2.0 or a new Business Model for IT Provisioning?, pp. 2–26
14. Youseff, L., Butrico, M., Da Silva, D.: Toward a Unified Ontology of Cloud Computing. In: Grid Computing Environments Workshop, GCE 2008, pp. 1–10 (2008)

15. Schweiger, B., Ehnert, P., Schlichter, J.: Simulative Evaluation of the Potential of Car2X-Communication in Terms of Efficiency. In: Strang, T., Festag, A., Vinel, A., Mehmood, R., Rico Garcia, C., Röckl, M. (eds.) Nets4Trains/Nets4Cars 2011. LNCS, vol. 6596, pp. 155–164. Springer, Heidelberg (2011)
16. Mehmood, R., Nekovee, M.: Vehicular Ad hoc and Grid Networks: Discussion, Design and Evaluation. In: Proc. of the 14th World Congress on Intelligent Transport Systems, p. 8 (2007)
17. Mehmood, R., Crowcroft, J., Hand, S., Smith, S.: Grid-Level Computing Needs Pervasive Debugging. In: The 6th IEEE/ACM International Workshop on Grid Computing, Seattle, WA, USA, pp. 186–193 (2005)
18. Rao, S.V.R.K., Diwanji, V.: WiMax'ble Pervasive Cloud – Empowering Next Generation Intelligent Railway Infrastructure. In: Strang, T., Festag, A., Vinel, A., Mehmood, R., Rico Garcia, C., Röckl, M. (eds.) Nets4Trains/Nets4Cars 2011. LNCS, vol. 6596, pp. 58–68. Springer, Heidelberg (2011)
19. Master Plan of the City of Al-Ramadi, Second Stage Report: Analysis of Existing Situation, Regional Context and Major Development Issues. Hydrosult Center for Engineering Planning (HCEP), Montreal, Canada (November 2009)
20. Mehmood, R.: Disk-based techniques for efficient solution of large markov chains. School of Computer Science, University of Birmingham, UK (2004)
21. Mehmood, R., Crowcroft, J.: Parallel Iterative Solution Method for Large Sparse Linear Equation Systems. Computer Laboratory: University of Cambridge (2005)
22. Mehmood, R.: Towards Understanding Intercity Traffic Interdependencies. In: Proc. of the 14th World Congress on Intelligent Transport Systems (2007)
23. Ayres, G., Mehmood, R.: On Discovering Road Traffic Information Using Virtual Reality Simulations. In: International Conference on Computer Modeling and Simulation, Los Alamitos, CA, USA, pp. 411–416 (2009)
24. Vegni, A.M., Little, T.D.C.: A Message Propagation Model for Hybrid Vehicular Communication Protocols. In: Proc. of IEEE 2nd Intl. Workshop on Communication Technologies for Vehicles (Nets4Cars 2010), Newcastle, UK, July 21-23 (2010)
25. Mejri, N., Kamoun, F., Filali, F.: Cooperative infrastructure discovery through V2X communication. In: Proc. of the 9th IFIP Annual Mediterranean Ad Hoc Networking Workshop, pp. 1–8 (June 2010)

Wireless Driver and Vehicle Surveillance System Based on IEEE 802.11 Networks

Krzysztof Różanowski[1], Zbigniew Piotrowski[2], Tadeusz Sondej[2], Krzysztof Sawicki[2], and Marcin Głowacki[2]

[1] Military Institute of Aviation Medicine, 54/56 Krasińskiego str., 01-755 Warsaw, Poland
[2] Military University of Technology, Faculty of Electronics, 2 Kaliskiego str., 00-908 Warsaw, Poland

Abstract. The paper presents a complex system for driver and vehicle surveillance. System uses many different sensors such as acceleration sensors, ECG and blood pressure measuring devices, eye movement recorders, video cameras, GPS and wireless IEEE 802.11 networks for transmission of recorded data in real time. The paper presents also two experiments with their results and a method to calculate a metric characterizing the driver's technique that uses signals recorded during the experiment.

Keywords: driving technique, wireless networks, IEEE 802.11, accelerations, gyroscopes, eye movement.

1 Introduction

Nowadays, cars are an integral part of our everyday life. Every year car manufacturers introduce more and more sophisticated vehicles, featuring more and more advanced driver and passenger safety systems. Still, it is the human who is responsible for driving the vehicle, and it is in human nature to be imperfect, to make mistakes, or to be inexperienced or simply tired. So far, no system designed to monitor the psychological and physical condition of the driver has been made commonly available. Nor do we have a mechanism designed to assess drivers' driving skills or technique. This publication presents the method of numerical assessment of the driving technique, considering the way the driver controls the vehicle, as well as some of the fatigue indicators.

2 Related Works

There are several papers discussing the field of monitoring the psychological and physical condition of vehicle drivers, as well as the technical condition and driving performance of the vehicle itself. Some papers contain data analyzed in real time, with the on-board devices installed in the vehicle. Other papers analyze data with the use of external devices, after completing the comprehensive measurement procedure.

A. Vinel et al. (Eds.): Nets4Cars/Nets4Trains 2012, LNCS 7266, pp. 57–67, 2012.

In [1] authors consider using a network of sensors located around the driver, functioning as detectors sensing the condition of individual vehicle components, in order to improve road safety. Also, the sensor system monitors the space of the radius of 15 ft (4.57 m) around the vehicle. The system is based on the TinyOs operating system and the TinyDb database. The sensors are based on the ZigBee technology. Article [2] focuses on detection of driver's drowsiness based on the DSP-based image processing unit. It highlights the option to establish communication with the external server via the 3G network. The drowsiness level is defined on the basis of the PERCLOS (Percentage of Eye Closure) measure. In [3], great emphasis is placed on remote monitoring of the vehicle and the driver. The system retrieves vehicle data via ODB (On-Board Diagnostics) and sensors (GPS, accelerometers), as well as cameras monitoring the driver and the road. The data is transferred to the remote server and made available via HTTP. Authors of [4] suggest a system for remote monitoring of senior drivers' health. It is assumed that a senior person is equipped with a device monitoring the person's health, which communicates with a transmitter (e.g. a smartphone) via Bluetooth. Next, data is transmitted via the mobile phone network to a remote server. The article focuses mainly on the method of data transmission, rather than data acquisition. [5] presents the method of transport fleet management and monitoring, which allows for analyzing the condition of the vehicle, as well as the driver. Vehicle data is read via the OBD interface, the vehicle position data is retrieved from the GPS system, and the driver is authenticated on the basis of the driver ID, via a RFID reader device. Data is transferred among the vehicles via Wi-Fi or a mobile phone network (HSPA+). In [6], with the use of a photoplethysmogram (PPG), the heart rate signal (HR) and the heart rate variability (HRV) signal is defined and analyzed on an ongoing basis. The PPG retrieval module communicates with a mobile device (e.g. a smartphone), via Bluetooth. Special software is used to analyze the signal, and if any abnormalities are detected, it alerts the driver and transmits the data via the mobile phone network into a remote server. Moreover, the software calls for help automatically. In [7], authors put forward a non-intrusive method of driver fatigue monitoring, based on analyzing the images captured by two cameras, as well as the PERCLOS measure calculated based on these images. Data acquisition and analysis is carried out on board of the vehicle. Authors of [8] define the level of safety while driving a vehicle, based on the driver and vehicle data. The capacity to drive the vehicle, as well as the vehicle's condition/responses, are regarded as factors supplementary to each other. Data is analyzed in real time, with the use of an on-board device. Analogically, in [9] the level of driver's fatigue is assessed based on the PERCLOS measure, degree of mouth openness, or head inclination angle. The difference is that the data is analyzed with the FPGA circuit, and the results of the analysis carried out on the external server are transmitted via GPRS. [10] puts forward a "ubiquitous" health information acquisition system. Electrodes have been placed on the steering wheel and door handles, in a way that does not compromise driving experience. The system records ECG and HRV signals, impedance of the human body, as well as blood pressure. Authors of [11] place most emphasis on vehicle monitoring, including the distance to objects in the space around the vehicle, as well as the vehicle's position and speed. This data is used for calculating the threat coefficient. After exceeding the threshold value, the coefficient triggers a driver alert, and transmits information to the remote monitoring center via a wireless module. Document [12] is an overview of driver inattention monitoring systems for intelligent vehicles and contains a comparison of most of the methods in use since the beginning of 2010. In [13] authors present a design of a built-in system based on the GNU/Linux platform that uses the 3G network for communication purposes. The data acquisition module collects standard ECG, blood pressure signals, etc. The same applies to [14], but the document focuses more on the device itself, functioning as a unique medical package, based on a GNU/Linux system and communicating with the server via the 3G network, as well. In [16] authors present their own, custom driver alertness surveillance system. As the upgrade of all the previous systems, this system utilizes electrodes used for EEG monitoring. The analyzed vehicle parameters are similar as in [2, 4, 5]. Authors of [17] attempt to define a vehicle driving model, with the use of a graph composed of 5 statuses (training another vehicle, braking resulting from the preceding car's braking, free driving, braking against traffic lights, launching against the green light) and transitions among them. Next, with the use of statistical analysis, they defined a coefficient that allows for recognizing the current state of the driver.

This paper presents a new approach to the problem of wireless system for driver and vehicle surveillance. In addition it presents a new method for numerical evaluation of driving techniques.

3 Experiments

Two experiments were carried out. 15 drivers, as well as a vehicle equipped with a number of sensors and recording devices, took part in the first experiment. Each driver took a 400 km test drive twice. The drives took place on Mondays and Fridays (fist day of the working week, in the morning, and the last day of the working week, in the afternoon). During each drive, the following signals were monitored and recorded: acceleration and deflection along the three axes, geographical position from the GPS, temperature inside the vehicle and the atmospheric pressure. Also, the system recorded biological signals: ECG, heart rate, blood pressure, skin temperature, skin impedance and blood saturation. Also, the vehicle was equipped with the FaceLAB v5 system, designed to record eye behavior parameters (PERCLOS, presence of saccades, etc.) Two cameras added up to a complete setup: one recorded the driver's face, the other one recorded the traffic conditions in front of the vehicle. All the data recorded during test drives were saved locally (e.g. onto SD memory cards) by recording devices. The recorded signals were read only after the test drive was finished. The equipment of laboratory vehicle is presented on fig. 1.

The other experiment was carried out as the follow up of the previous experiment, enhanced with the feature of wireless transmission of signals during test drives, so that it would be possible to carry out the analysis on an ongoing basis, e.g. by the vehicle driven close to the analyzed vehicle.

Fig. 1. Equipment of the laboratory vehicle

3.1 LoggerBox

During the second experiment a device called LoggerBox was used. The device is available as a small box, with the following modules installed inside it: a GPS module for setting the geographical position, a module measuring deflections along the three axes, a module measuring acceleration along the three axes, together with the following modules: a wireless communication module compliant with 802.11 b/g with the full TCP/IP stack, and the USB wired communication module. In addition, the device is equipped with a SD memory card, which allows for storing the measurement session data. Since the device is equipped with high sensitivity acceleration meters, it should be fixed so that it cannot move freely during test drives, which prevents the device from moving inside the vehicle and thus does not invoke data recording errors. The device is designed to be powered up via the vehicle's battery. It also features the CAN bus, which allows for plugging in a number of other devices. Ultimately, LoggerBox will be used for plugging in all devices used in the first experiment (or their alternatives, recording the same parameters and signals), additional modules recording vehicle responses with the use of the on-board computer unit (engine rpm, fuel consumption, etc.), as well as devices recording the distance to the cars preceding the analyzed vehicle, and devices recording environmental parameters (such as noise, pollution, humidity, temperature). All the above-mentioned signals and parameters will be available for wireless transmission in real time.

LoggerBox is used with custom software developed in C# specifically to support the device. The software allows for remote retrieval of recorded measurement sessions from the device and remote monitoring of recorded signals in real time.

3.2 Experiment Preparations

The purpose of the experiment was to test LoggerBox functionality in real-world conditions. Two vehicles were used for the experiment. One of the vehicles was equipped with LoggerBox (fig. 2).

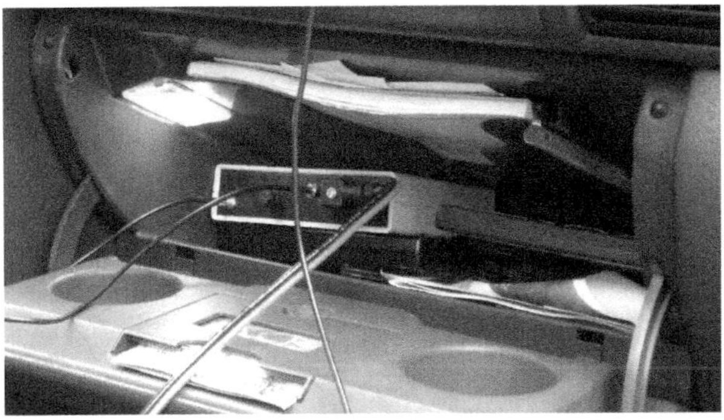

Fig. 2. LoggerBox fitted in the vehicle

The other vehicle performed the role of a data receiver and was equipped with a notebook with the external wireless network adapter, featuring a RP-SMA socket for plugging in the aerial. In order to ensure the best wireless connectivity parameters available, both vehicles were equipped with external omnidirectional aerials of 9 dBi gain. The wireless modules operated in the 802.11b/g ad-hoc (device-to-device) mode network, with no extra security measures enabled.

3.3 Experiment Execution

The experiment was carried out along a predefined route. The route was covered by both the vehicles. The vehicle equipped with LoggerBox was the first car in the convoy, followed by the monitoring vehicle. The distance between the vehicles was not constant and varied during the course of the entire experiment. The predefined route was divided into clearly identifiable sections, such as: straight sections, crossroads and bends. During the test drive, the monitoring vehicle used wireless connectivity to remotely record signals from the monitored vehicle.

During the test drive, on several occasions the monitoring vehicle purposefully increased the distance to the monitored vehicle, up to the point where it is impossible to establish a connection (approximately 150 m). Lack of connection rendered real-time signal recording impossible. However, all the signals were also recorded onto the memory card. Therefore, as soon as the connection was reestablished, it was possible to record the missing signals by downloading them from the device memory.

3.4 Acceleration Analysis

Analysis of recorded acceleration signals gives the opportunity to restore the route of the vehicle. Clearly visible are some specific situations such as braking, accelerating, turning, etc. On the basis of the maximum values recorder during the test runs, it is possible to specify characteristics such as acceleration and braking of the car. Determination of values exceeded maximal recorded values may mean the occurence of collisions and accidents. Examples of acceleration signals are shown in fig. 3-6.

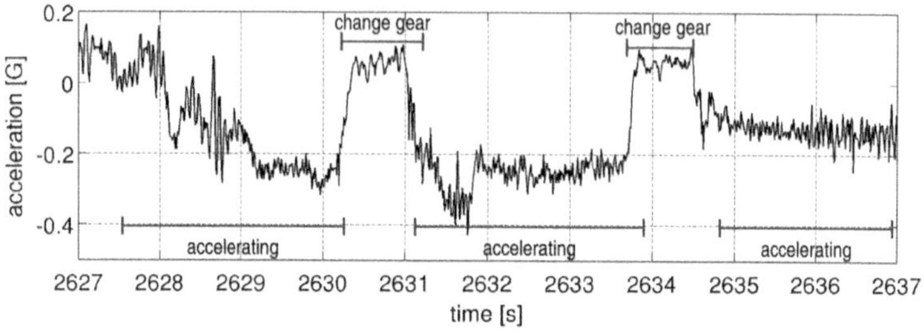

Fig. 3. Gear changes recorded in acceleration signal on X axis

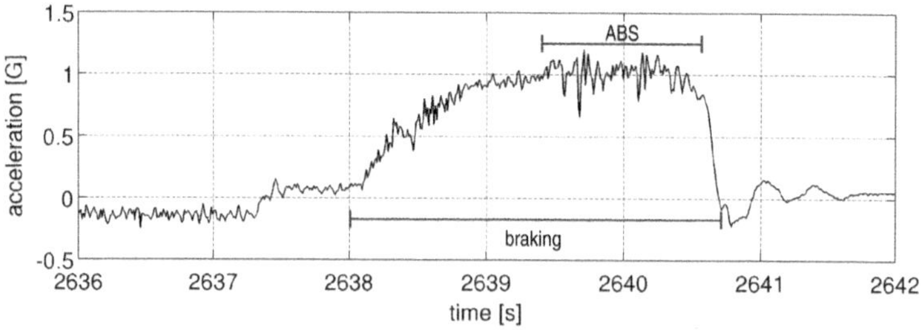

Fig. 4. Braking recorded in acceleration signal on X axis

4 Driver's Driving Technique

Data gathered during the experiments were analyzed to calculate a numerical value describing the driver's driving technique. Thus, the route for each test drive was divided into characteristic sections. As regards the first experiment, over 500 sections were defined, such as: speed limits, bends, long straight sections and crossroads. Speed limit sections are the largest group of characteristic sections. Thus, it was necessary to develop a measure allowing for describing the driver's tendency to speed. To do so, a SOLP (Speed Over Limit Parameter) coefficient was defined, calculated for each speed limit section:

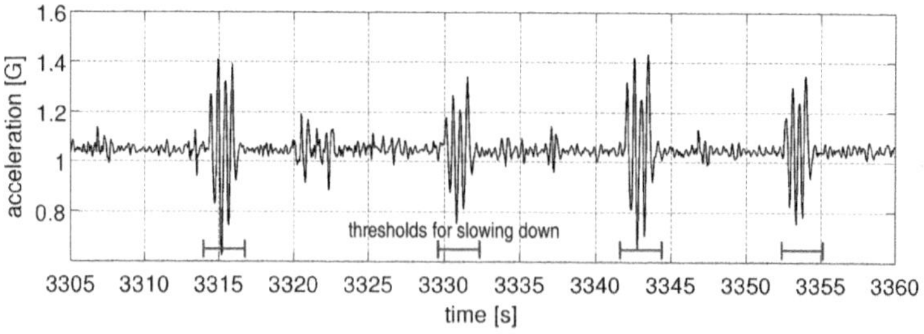

Fig. 5. Driving through slowing down thresholds recorded in acceleration signal on Z axis

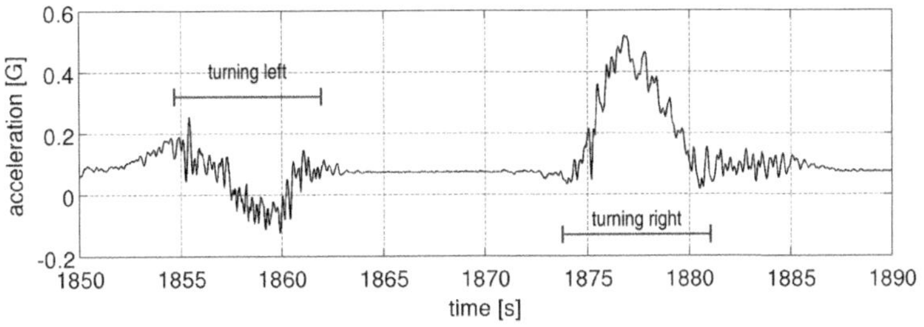

Fig. 6. Turning right and left recorded in Y axis

$$SOLP = \frac{\sum_{n} \begin{cases} (v_n - v_l) & if\ v_n > v_l \\ 0 & otherwise \end{cases}}{n} \tag{1}$$

where n is the number of speed signal samples, v_n – value of the speed signal sample, and v_l – speed limit value.

More detailed data analyses involved selecting the most significant signals for the driver's driving technique. After an initial analysis of the collected data quality, signals of accelerations along the three axes, the PERCLOS signal and the EyeClose signal for the degree of eyelid closure, measured in the 1-second variable window, were selected for further processing. The reason for using the extra EyeClose signal is that PERCLOS counts only the cases when the percentage of eye closure exceeds 80%. This signal is in the range of 0-1, where 0 stands for the eye fully open, while 1 stands for the eye fully closed for the entire measurement window duration. Moreover, the following parameters have been calculated: the shortest detected time between blinks and the shortest detected time between saccades.

The NMSE (Normalized Mean Square Error) metric was calculated for the signals and parameters:

$$NMSE = \frac{\sum_{n} (A_n - A')^2}{\sum_{n} A_n^2} \tag{2}$$

These metrics allowed for developing the ROAD coefficient, describing the driver's driving technique and allowing for classifying the driver into one of the three groups: safe drivers, normal drivers, aggressive drivers.

$$ROAD = EyeParams + AccParams + 0.5 \cdot SOLP \tag{3}$$

$$\begin{aligned} EyeParams = 0.1 \cdot EyeClose + 0.1 \cdot MinBlinkDur_{NMSE} + \\ + 0.1 \cdot MinSaccadeDur_{NMSE} + 0.1 \cdot PERCLOS_{NMSE} \end{aligned} \tag{4}$$

$$AccParams = 0.2 \cdot AccX_{NMSE} + 0.05 \cdot AccY_{NMSE} + 0.05 \cdot AccZ_{NMSE} \tag{5}$$

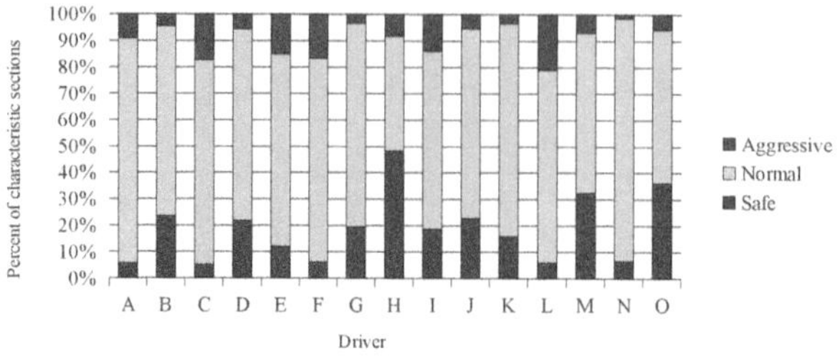

Fig. 7. Percentage share of driving techniques for individual drivers

The norm values used for calculating the NMSE metric for signals were as follows: MinBlinkDur' = 3 s, MinSaccadeDur' = 0.25 s, PERCLOS' = 0,2 %, AccX = 0 G, AccY = 0 G, AccZ = 1 G. The SOLP and EyeClose parameters are used in the unmodified form. Most values for this coefficient were within the 0-0.5 range. Values exceeding 0.5 were present on rare occasions, mostly involving unexpected, emergency situations on the road.

The ROAD coefficient values allow for classifying the driver to one of the groups (safe, normal, aggressive drivers). In order to define the group limit thresholds, it was necessary to carry out the subjective evaluation of test drives, based on the video recordings of selected test drives. It was decided that the ROAD coefficient values of 0-0.05 classify the drivers into the safe driver group, 0.05-0.15 classify them into the normal driver group, and values exceeding 0.15 classify the drivers into the aggressive driver group. Because the ROAD coefficient was calculated for each of the characteristic sections, it was possible to use the histograms of these values to identify the percent of characteristic sections covered by drivers in an aggressive or normal manner (Fig. 7).

5 Environmental Conditions

During test drives, signals for atmospheric pressure and vehicle interior temperature were acquired. The atmospheric pressure is an essential parameter, especially when considering persons susceptible to any atmospheric impacts (weather-sensitive persons). As outlined in [15], the change of the atmospheric pressure is sensed, if it exceeds 4 hPa per hour. For environmental conditions, a coefficient was developed, displaying their quantitative measure. The generic coefficient formula is presented with item (6). This coefficient (ENV) is the sum of partial coefficient values (currently they include pressure – ENVpressure and temperature – ENVtemp), with the appropriate weight divided by the sum of weights.

$$ENV = \frac{\sum_{i=1}^{N} a_i \cdot ENV_i}{\sum_{i=1}^{N} a_i} \tag{6}$$

The first of its components is the component responsible for atmospheric pressure changes ($ENV_{pressure}$) exceeding the specific threshold for a given time window.

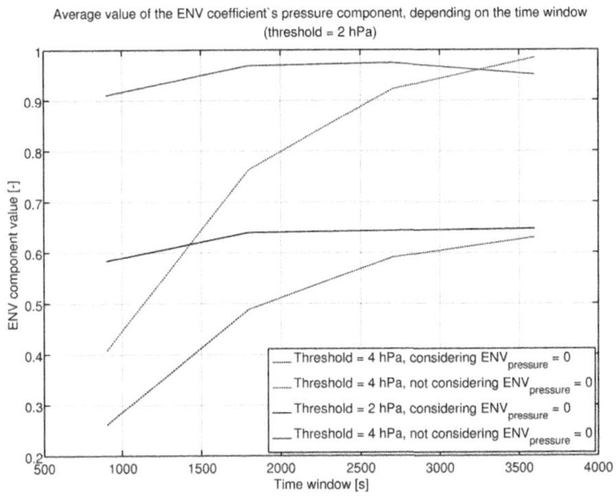

Fig. 8. Average values of environmental coefficient's pressure component

The values for the pressure component are presented in the graph (fig. 8) for various time windows (900s - 3600 s) and various thresholds (2 hPa or 4 hPa). As indicated by the tests, the average coefficient value was above 0.8 (over 80% of the test drive has a potentially negative impact on the driver's state of mind). Some cases were observed, where the coefficient value was close to 0, usually present when driving along short route sections. In order to differentiate the impact of these values onto the average value, two patterns were defined – with and without the $ENV_{pressure} = 0$.

The optimum temperature inside the passenger cabin should be between 20°C to 22°C. Based on this assumption, a module for calculating the coefficient responsible for the impact of temperature on the driver was developed. Average coefficient values depending on the assumed threshold (from 20°C to 33°C) are presented in the graph (fig. 9) (the value closer to 0 is the desirable one). The results indicate that most frequently the drivers travelled in the temperature of 26 – 28°C (the largest coefficient value drop).

Ultimately, the environmental coefficient will be enhanced with additional signals, such as the driver face illumination degree, noise level, pollution level or air humidity.

6 Wireless Transmission

During the wireless transmission experiment, on several occasions it was possible to establish wireless connection between the two vehicles moving along the route, and read the signals recorded in real time on board of the first vehicle. Next, the recorded signals were analyzed. The driver's driving technique was analyzed with the use of a simplified $ROAD_{Acc}$ coefficient, composed only of the acceleration signals, and defined in the following way:

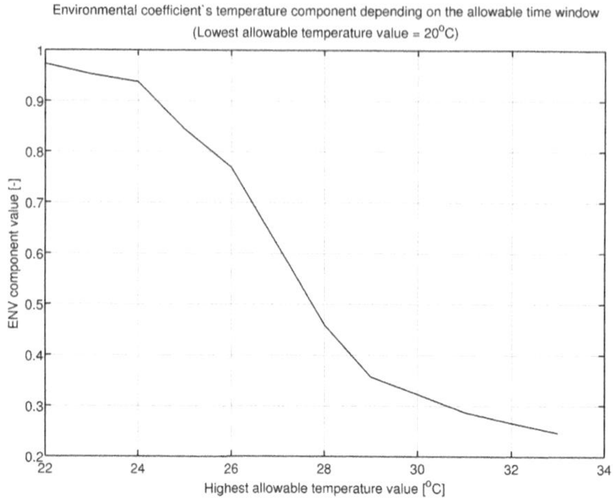

Fig. 9. Environmental coefficient's temperature component

$$ROAD_{Acc} = 0.5 \cdot AccX_{NMSE} + 0.3 \cdot AccY_{NMSE} + 0.2 \cdot AccZ_{NMSE} \quad (7)$$

This coefficient was calculated for the entire route covered by the vehicle. The coefficient calculated for the entire test drive was 0.022, which meant that the driver was classified into the safe driver group.

7 Summary

The experiment results prove it is possible to carry out remote and wireless monitoring of signals recorded in the monitored vehicle. Defining driver's driving profiles in real time may prove extremely useful for the sake of courses for professional drivers. Upgrading the system with the option designed to record vehicle signals (engine rpms, etc.) will allow for developing an all-new system, capable of assessing the driving style in real time. The system concept may be also used for spotting drivers violating traffic regulations. A device functioning as the "accident data recorder", installed onboard of the vehicle, may communicate with Police cars, so that Police officers are able to track the driver's behavior in order to detect possible traffic regulation violations.

Acknowledgment. The results presented herein have been achieved within the framework of the project "An integrated system for monitoring the psychophysiological state of vehicle drivers to minimize threats in road traffic", project No. UDA-POIG.01.03.01-10-085/09.

References

1. Chin, H.H., Jafari, A.A., Hwang, Y.-F.: Solutions of Wireless Sensor Networks for Driver Associate. In: 40th Southeastern Symposium, pp. 356–361 (2008)

2. Weng, M.-C., Chen, C.-T., Kao, H.-C.: Remote Surveillance System for Driver Drowsiness in Real-time Using Low-cost Embedded Platform. In: Vehicular Electronics and Safety, pp. 288–292 (2008)
3. Fu, X., Men, Y., Yuan, G.: A Multimedia Database Application for Vehicle Surveillance System. In: 2010 2nd International Workshop on Database Technology and Applications, pp. 1–4 (2010)
4. Yamagishi, H., Suzuki, H., Watanabe, A.: Study of a Remote Monitoring System for Senior Drivers. In: TENCON 2010 - 2010 IEEE Region 10 Conference, pp. 1042–1047 (2010)
5. Aljaafreh, A., Khalel, M., Al-Fraheed, I., Almarahleh, K., Al-Shwaabkeh, R., Al-Etawi, S., Shaqareen, W.: Vehicular Data Acquisition System for Fleet Management Automation. In: 2011 IEEE International Conference on Vehicular Electronics and Safety, pp. 130–133 (2011)
6. Lin, Y.-H., Lin, C.-F., Wu, C.-R., Hsu, Y.-C., You, H.-Z.: Intelligent Physiological Monitoring and Feedback System for Driver Safety. In: 2011 Fifth International Conference on Genetic and Evolutionary Computing, pp. 25–28 (2011)
7. Zhu, Z., Ji, Q.: Real Time and Non-intrusive Driver Fatigue Monitoring. In: The 7th International IEEE Conference on Intelligent Transportation Systems, pp. 657–662 (2004)
8. Mueller, F., Wenzel, A.: Intelligent Monitoring and Adaptive Competence Assignment for Driver and Vehicle. In: Intelligent Vehicles Symposium, pp. 1101–1106 (2007)
9. Dai, J., Zhang, J.: Information Fusion System with Wireless Transmission for Monitoring the Status of Driver. In: Industrial Electronics and Applications, pp. 2670–2673 (2007)
10. Jeong, C., Jun, S.H., Lee, D.H., Yoon, H.R.: Development of Bio Signal Measurement System for Vehicles. In: Convergence Information Technology, pp. 1091–1096 (2007)
11. Lee, T.-T., Chiang, H.-H., Perng, J.-W., Jiang, J.-J., Wu, B.-F.: Multi-sensor Information Integration on DSP Platform for Vehicle Navigation Safety and Driving Aid. In: Networking, Sensing and Control, pp. 653–658 (2009)
12. Dong, Y., Hu, Z., Uchimura, K., Murayama, N.: Driver Inattention Monitoring System for Intelligent Vehicles: A Review. In: Intelligent Vehicles Symposium, pp. 875–880 (2009)
13. Peng, Z., Hu, C., Liao, J., Chang, T.: Design of Embedded Remote Monitoring Terminal Based on 3G Network. In: Complex Medical Engineering, pp. 274–278 (2011)
14. Li, P., Liao, J., Hu, C., Yin, Y., Meng, M.Q.-H.: Design of Medical Remote Monitoring System Base on Embedded Linux. In: Information and Automation, pp. 590–594 (2011)
15. Pilcher, J.J., Nadler, E., Busch, C.: Effects of hot and cold temperature exposure on performance: a meta-analytic review. Ergonomics 45(10), 682–698 (2002)
16. Portouli, E., Bekiaris, E., Papakostopoulos, V., Maglaveras, N.: On-road experiment for collecting driving behavioural data of sleepy drivers. Somnologie 11, 259–267 (2007)
17. Khaisongkram, W., Raksincharoensak, P., Shimosaka, M., Mori, T., Sato, T., Nagai, M.: Automobile Driving Behavior Recognition Using Boosting Sequential Labeling Method for Adaptive Driver Assistance Systems. In: Dengel, A.R., Berns, K., Breuel, T.M., Bomarius, F., Roth-Berghofer, T.R. (eds.) KI 2008. LNCS (LNAI), vol. 5243, pp. 103–110. Springer, Heidelberg (2008)

Architecture of Car Measurement System for Driver Monitoring

Krzysztof Różanowski[1], Tadeusz Sondej[2], Zbigniew Piotrowski[2], and Krzysztof Sawicki[2]

[1] Military Institute of Aviation Medicine, Facility of Aviation Engineering, 54/56 Krasińskiego str., 01-755 Warsaw, Poland
[2] Military University of Technology, Faculty of Electronics, 2 Kaliskiego str., 00-908 Warsaw, Poland

Abstract. This paper addresses the today's social needs to ensure security in road traffic, a task which becomes difficult due to the growing number of drivers and, therefore, increasing traffic intensity. The rising need for mobility constitutes a huge challenge for the institutions responsible for improving the road traffic safety. The paper describes selected hardware components which can be used to build an integrated warning system. Its measuring modules supply data on the road traffic environment parameters, the vehicle dynamics, and psychophysiological behavior of the drivers. An important aspect of interoperation of such hardware components is reliable communication and data synchronization. The presented solutions are based on the CAN, WiFi, ISM 2.4 GHz, and GSM/GPRS interfaces. To minimize the risk of data loss, the data is stored redundantly in the local memory structures, as well as in a remote archiving system to which the data is transmitted through wireless communication protocols.

Keywords: measurement system, driver monitoring, data processing, embedded system, data logger, dynamic modeling, road safety.

1 Introduction

The concept of developing the measuring components described below emerged in result of an attempt to view the road-user safety in a new way through explaining and describing the phenomena connected with the drivers' movements in the road traffic space as multiple component models build on the basis of data describing the driver's psychophysiological capacity and predisposition, the influence of external factors, the technical parameters of the vehicle, and the distractive effect of the environment. The Polish road safety reports show that in 2010 there were 38.832 road accidents with 3.907 people killed and 48.952 injured. The human is definitely the key factor among numerous circumstances contributing to the accident, such as the season, time of the day, illumination, surface type, weather conditions, and vehicle speed. The decisions made by the driver with respect to the driven vehicle constituted the critical cause in nearly

A. Vinel et al. (Eds.): Nets4Cars/Nets4Trains 2012, LNCS 7266, pp. 68–79, 2012.

80% of accidents in 2010. In result of wrong human decisions, 2.633 people were killed and 40.711 were injured last year [1].

From the traffic-engineering point of view, the factors influencing the road traffic can be classified into three main groups: vehicles (such as passenger cars, trucks, slow-moving vehicles, streetcars, one-track vehicles), road users (such as drivers, passengers, cyclists, pedestrians), and road environment (including the road with its engineering structures, devices, and installations, the road surroundings, traffic conditions, weather conditions). All those factors constitute an interworked network, contributing in various degrees to the occurrence of certain road events and situations.

Measuring simultaneously all factors affecting the driver's behavior and at the same time the driver's psychophysiological state is a difficult task[13,14]. The available literature contains descriptions of some methods and devices, but typically they support only selected parameters [2,3,5-7,10,11,15-18]. In software scope, the proposed design and standardization of the firmware of in-car sensors has been presented in [4,9]. Therefore, works have been undertaken to build a system supporting a comprehensive set of parameters.

This paper presents selected hardware components which, as the objective, will enable building a system supporting tests to identify the psychophysiological indicators responsible for the fatigue level and for deterioration of the situation awareness and psychomotor skills which is an unique approach. Collecting the data about factors affecting the fatigue and skill levels may be useful for developing in the future certain prevention programs counteracting the drivers' fatigue. Also, analyzing and comparing the results of laboratory tests with results of tests conducted during actual driving will enable evaluation of the predictive value and usefulness of traditional tools used for testing the drivers. The concept of developing a unique, mobile diagnostic apparatus, presented in this paper, will enable the tests to be conducted in laboratory conditions, in simulated conditions, and in real traffic conditions.

2 System Organization Chart for Driver Monitoring

The measuring system consists of multiple hardware, software, and organizational elements. Some of them are tightly interconnected and other ones can work autonomously. Furthermore, some elements are optional, depending on the expected configuration. Two basic configurations, or modes, can be distinguished: on-line (with real-time data monitoring) and off-line (with data processing after the test completion). Data post-processing is possible also in the on-line mode, because all data and signal samples are stored in the recorder memory. Figure 1 below shows a block diagram of the whole measuring system.

The basic measuring component is referred to as the Car Measurement System (CMS). Its purpose is to monitor the driver's behavior and psychophysiological state, to measure the vehicle movement parameters, and to monitor the road situation and environmental parameters inside and outside the vehicle. All those parameters should reconstruct, as accurately as possible, the conditions and

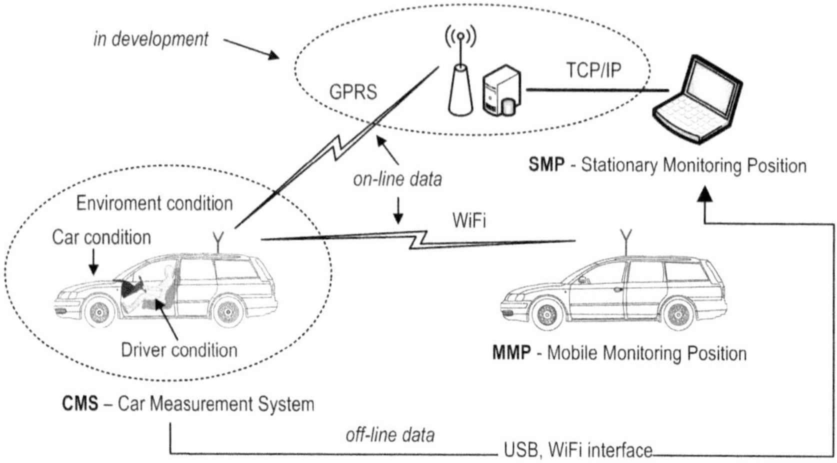

Fig. 1. Block diagram of the system

factors influencing the driver and thus enable studying the driver's behavior. Each measured and processed signal is stored in non-volatile memory of CMS and can be later read for the purpose of off-line data processing, analysis, and visualization. The data can be read through the USB or WiFi interface. Also, the signals measured and processed in CMS can be transmitted in real time to the monitoring workstation Which can be either mobile (MMP on the figure) or stationary (SMP). The mobile workstation must be installed in a vehicle traveling closely behind the monitored vehicle. If external WiFi antennas are used, the communication range is a few hundred meters. To ensure maximal comfort of the driver under test (unawareness of the MMP vehicle in the vicinity), as well as to monitor multiple vehicles distributed over a large area, the system envisages real-time data transmission through the GSM/GPRS network. This variant is currently being deployed.

3 Architecture of Car Measurement System

3.1 Block Diagram of CMS

CMS is the main component responsible for controlling, data acquisition, and pre-processing. It performs all functions related to configuring the system for operation, reads the sampled signals from sensor modules, performs the necessary data processing, writes the data to non-volatile memory, and is responsible for communication with the monitoring workstation. CMS is a complex, distributed measuring system consisting of several microprocessors (microcontrollers). The CMS system architecture is shown on Fig. 2 below.

CMS includes elements installed both inside and outside the vehicle. One of them, PMP (Personal Medical Packet) is a personal device worn by the driver on

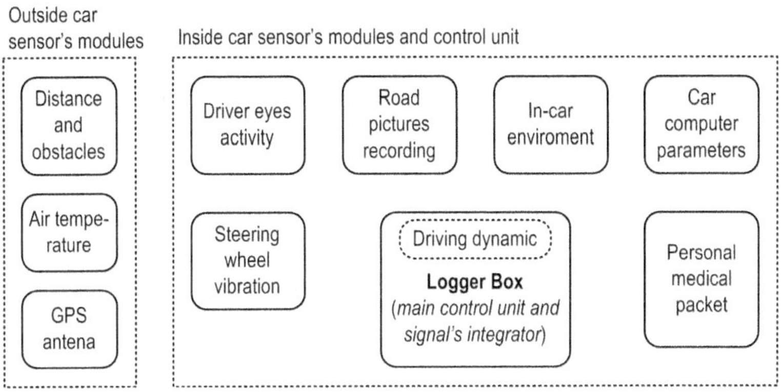

Fig. 2. CMS block diagram

his/her body at all times. It enables the driver's behavior to be monitored also during breaks in driving and thus ensures continuous recording even when the driver temporarily leaves the vehicle. CMS is a modular system. The following modules are installed outside the vehicle:

- Sensor module for detecting obstacles and measuring their distances; it determines the distance and angular direction to the obstacle. In includes dedicated radars installed at the front and rear of the vehicle.
- Ambient temperature sensor module.
- Optionally, a GPS antenna to improve the transmitted signal quality.
- The following modules are installed inside the vehicle:
- Module responsible for oculomotoric activity detection and road image recording. The module detects when the driver closes his/her eyes, by analyzing the video image, taking advantage of the bright eye phenomenon. Thanks to such solution, no special equipment needs to be installed on the driver's head. Optionally, the module records the road image in front of the car and the image of the driver, using two cameras.
- Module for measuring the steering wheel vibrations and turning angle. It uses accelerometers and gyroscopes.
- Module for measuring the vehicle dynamics, determining the vehicle location, acceleration, and deceleration. It uses a GPS receiver, accelerometers, and gyroscopes.
- Module measuring the environmental parameters inside the vehicle and illumination from ahead. It determines a number of factors affecting the driver's comfort, such as air temperature, humidity, air pressure, pollution, and noise. Importantly, it also measures illumination of the driver's eyes, which is particularly significant during driving in high insolation or at night.
- A module responsible for reading data from the onboard computer, such as traveling speed and engine rotational speed, through the OBD II (On-Board Diagnostic) information system.

- PMP (Personal Medical Packet), measuring the psychophysiological parameters and motoric activity of the driver. It measures the ECG signal (to determine the heart rate (HR) and its variability (HRV)), body temperature, blood pressure, atmospheric pressure, and body movements, using accelerometers and gyroscopes.
- Logger Box — the main control unit. In integrates all measuring modules, pre-processes the data, and communicates the with monitoring workstation.

3.2 Measuring Module Design

CMS consists of distributed measuring modules. Each of them follows a similar design shown on Fig. 3 below.

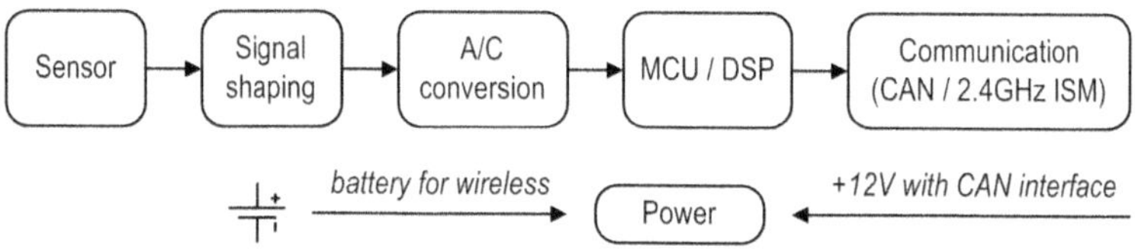

Fig. 3. Typical design of a sensor module

An important aspect if each module is the type of the used sensor, which determines the nature and quality of the measured signal. An input analog signal is fed to the shaping circuit (e.g. filters, level adjustment circuits) and then to an analog/digital (A/D) converter. The converter is selected to provide the necessary conversion resolution, accuracy, and speed. The sampling and necessary data preprocessing is performed by the microcontroller or specialized DSP processor. The measuring module communicates with the Logger Box through the CAN interface or wirelessly in the ISM 2.4 GHz band. A module with wire connection is powered through the CAN bus (CAN interface with an additional power cable) and a module with wireless connection is powered from its own battery.

3.3 Communication Architecture

The communication architecture of CMS includes a number of digital interfaces, both wire-based and wireless. The interface type corresponds to the requirements in terms of transmission speed, resistance to interference, and interoperation with other interfaces. Fig. 4 below shows a block diagram of the communication system.

Since the measuring modules are installed at various places inside and outside the vehicle, they are connected with the Logger Box through the CAN interface. The CAN interface ensures sufficient transmission speed and resistance to interference. It also supports the node mode of operation, which enables additional

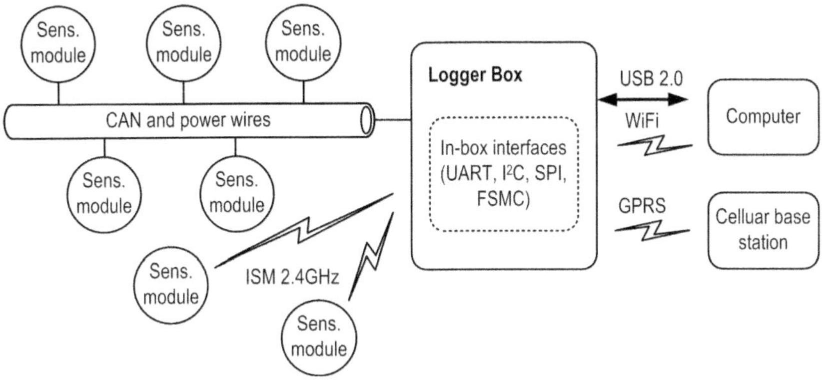

Fig. 4. Organization of CMS communications

sensors to be connected to the system. Each connected node (measuring module) has its ID, name, and operating mode register. The CAN nodes are slave devices to the Logger Box which is always the master device. The communication is invariably based on the "master's inquiry - slave's response" principle. The response from a slave device should occur within 200 μs. Such organization ensures synchronous sampling of all signals and continuous checking of the connection with the measuring module. In addition to the CAN interface signals (CANL and CANH), power cables (GND, +12 V DC) are connected to each module.

Measuring modules which cannot use wire connections communicate through wireless links operating in the ISM 2.4 GHz band. For that purpose, proprietary radio modules operating in the master-multislave configuration have been developed. They ensure very short connection establishing time (< 1 ms) and short time of slave's response to master's inquiries (< 4 ms).

The Logger Box communicates with the monitoring workstation through the USB 2.0 or WiFi interface. Those interfaces can be used interchangeably. Also, the Logger Box is equipped with a GPRS modem for communication with the cellular base station. In addition to external interfaces, the Logger Box uses the following interfaces for local transmissions between the microprocessor system elements: UART (Universal Asynchronous Receiver and Transmitter), I^2C (Inter-Integrated Circuit), SPI (Serial Peripheral Interface), FSMC (flexible static memory controller).

4 Data Processing Unit

4.1 Signals and Data Flow

The Car Measurement System architecture supports measuring more than 50 signals synchronously, in real time. The implementation of that task requires using complex control and sampling algorithms based on the TDMA (Time Division Multiple Access) technique. The signals can be processed both in real time and off-line. Due to the high complexity of the measuring system, it was necessary to clearly separate the controlling, data acquisition, and data processing functions. The data flow is shown on Fig. 5 below.

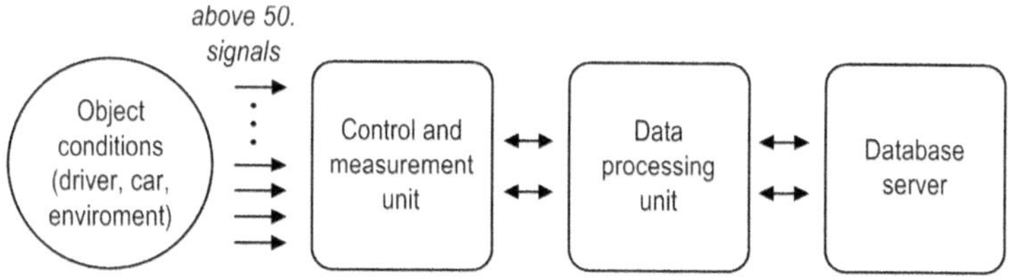

Fig. 5. Data flow within measurement systems

Signals from measured objects (the driver, vehicle, environment) are sampled by a programmable control and measurement unit. The samples are stored in the memory and may be transmitted to the data processing unit. Depending on the needs, the calculated values can be sent to the control and measurement unit or to the database server. Also, the data processing unit can access the database which contains calculated values or reference signals.

4.2 Data Processing

In CMS, it is assumed that the data recorded in the Data Logger module is analyzed with advanced signal processing algorithms. For that purpose, the data is converted to sets of factors reflecting the driver's health and driving predisposition. Those factors are listed in Table 1 below. For all listed factors, normalized mean square error is assumed as the metric to ensure that the obtained values always stay within the normalized range and therefore it is always possible to find the threshold values best representing the three basic categories of values: low, optimal, and elevated. Those categories are represented by scores from the following set: $\alpha = \{-1, 0, 1\}$.

The Automatic Data Processing System (ADPS), implemented both as a decision-making module in the Web server (for off-line processing) and in the Logger Box (for on-line processing), can make decisions regarding the driver's character, driving style, and driving predisposition. The on-line processing procedures implemented on the dedicated platform are written in the C language in the NetBeans IDE 7.0.1 environment on the Ubuntu GNU/Linux 11.10 operating platform.

The factors are calculated from multiple input parameters, through multiple iterations. The newest achievements in the area of biomedical signal analysis [12] have been used, including signal parameterization and classification procedures.

The Web-based decision-making module receives data recorded by CMS in the Data Logger and works out decisions regarding, inter alia, the subject's driving predisposition. For that purpose, it is necessary to establish decision thresholds on the basis of statistical tests on a large population of drivers. The decision-making

Table 1. List of factors calculated from signals generated by the Car Measurement System (CMS)

No	Factor	Description
1	ENV	Signals from the module measuring the environment conditions in the vehicle (such as air temperature and pressure)
2	ROAD	Signals from the module detecting oculomotoric activity and recording the road image
3	CAN	Signals from the module reading data from the vehicle's onboard computer
4	CARBIO	Parameters read from the onboard computer, correlated with parameters from the Personal Medical Packet (PMP)
5	BIO	Biological signals from the PMP

Table 2. A report generated by the decision-making module for the BIO factor (the value range for the BIO factor is {0-1})

RESULT FILE: ID30_252.res	
CmFlickerFact:	0.8871
CzFlickerFact:	0.8761
HrFact:	0.4803
ImpFact:	1.0000
SBPressFact:	0.6074
DBPressFact:	0.0996
Spo2Factor:	0.9577
TempFact:	0.7577
BIO:	0.7083

module provides a reporting panel and an administrative panel. Table 2 below shows an example report generated by ADPS for the BIO factor during our pilot tests.

where:

CmFlickerFact – flickering factor for the Flicker sensor
CzFlicker – decay factor for the Flicker sensor
HrFact – Heart Rate factor for the ECG signal
ImpFact – skin impedance factor
SBPressFact/DBPressFact – systolic/diastolic blood pressure factors
Spo2factor – blood saturation factor
TempFact – driver's body temperature factor.

In the case illustrated in Table 2, the calculated BIO factor value is 0.7083, classified to the decision range of $\alpha = 1$, i.e. the normal value. Works are under way to develop a set of rules and assessment criteria for the factors listed in

Table 1. The software modules will also be implemented on the Microsoft .NET development platform in order to optimize the code and thus to enable quasi-real-time simulation and experimental research.

5 Selected Testing Results

To verify the correct operation of the CMS hardware and software elements, several experimental tests have been conducted. Particular attention was paid to proper configuration and communication between the devices, as well as proper data synchronization and storage. During the tests, the equipment was installed in a car. Before starting the data recording, the system operation was checked by inspection of the data transmitted online through the WiFi interface. Then the test proper was conducted and after its completion the results were read through the USB interface. Example results are shown on Fig. 6 (the recorded route with characteristic points) and Fig. 7 (selected signals corresponding to the route).

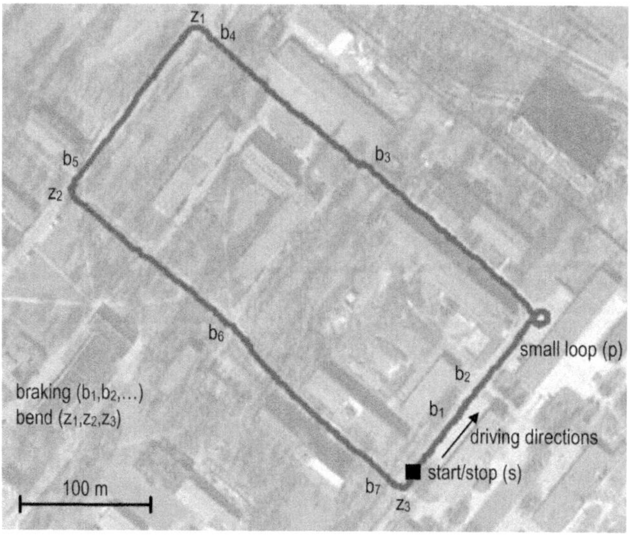

Fig. 6. The test ride route with characteristic points

To visualize the travelled route, the http://www.gpsvisualizer.com/ service has been used. The obtained photo has been marked with the following characteristic points: s - test start/end; p - loop with a diameter of ca. 5 m; $b_1 \ldots b_7$ - braking points; $z_1 \ldots z_3$ - turning points. The diagram based on the recorded GPS data truly reflects the travelled route, as visible e.g. for the loop.

The signals recorded during the ride have been read and visualized in the dedicated software application. The application enables viewing all or selected signals. Some of the signals are shown on Fig. 7 below.

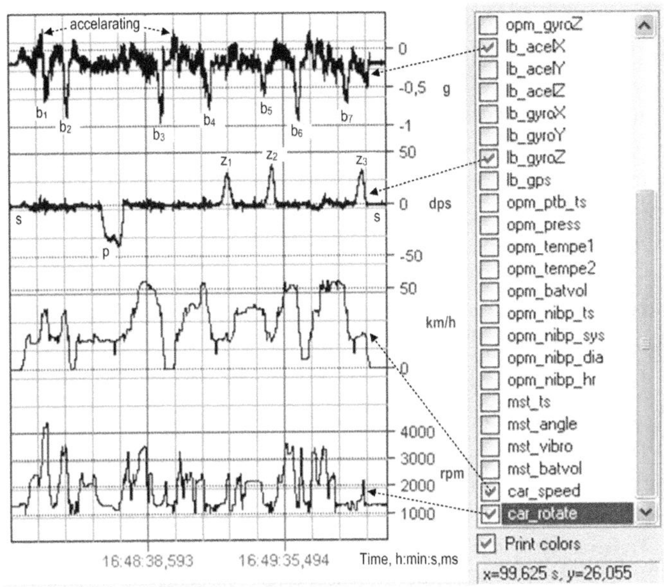

Fig. 7. Selected signals recorded for the route shown on Fig. 6

The figure presents four signals: *lb_acelX* — signal from the X-axis accelerometer of the dynamics measurement module; *lb_gyroZ* — signal from the Z-axis gyroscope of the dynamics measurement module (dps = degrees per second); *car_speed* — car speed; *car_rotate* — engine rotational speed (rpm = revolutions per minute). All signals are plotted in the function of time (shown in the hours:minutes:seconds,milliseconds format). The characteristic points, corresponding to those on Fig. 6, are marked. On the *lb_acelX* graph, the breaking points ($b_1 \ldots b_7$) are visible. At those points, *car_speed* was quickly decreasing and *car_rotate* was dropping to the idle value (ca. 1300 rpm). The *lb_acelX* signal shows also the car acceleration. On the *lb_gyroZ* graph, the moments of making the loop (p) and turns ($z_1 \ldots z_3$) are visible. The *lb_gyroZ* signal value is negative for right turns and positive for left turns.

6 Summary

The analysis of traffic accident statistics, available literature, and results of the pilot tests conducted by the authors confirms the importance of monitoring the drivers' psychophysiological state under distractive conditions. Such monitoring provides information which, as the ultimate goal of the research, will enable developing a warning system. The key component of such system will be a module evaluating the driver's predisposition, integrated with an elaborate sensor system. Such module is currently under development.

The example results shown above confirm the proper operation of CMS. Sufficient quality of the measured signals and their synchronous recording in real time have been obtained. The analysis of all (more than 50) signals may provide the full picture of the driving conditions and the driver's behavior.

The results obtained from the research conducted with the apparatus presented herein may be used to design future training programs in the area of road-situation awareness, as well as to train and rehabilitate those cognitive and psychomotor functions of people from various age groups, which deteriorate with age. In the future, the system based on the presented results may be used by other organizations and research centers developing programs aimed at increasing the security of road traffic, by healthcare entities specializing in rehabilitation of aged persons, and by driving schools which provide supplementary specialist training.

Acknowledgment. The results presented herein have been achieved within the framework of the project "An integrated system for monitoring the psychophysiological state of vehicle drivers to minimize threats in road traffic", project No. UDA-POIG.01.03.01-10-085/09.

References

1. Bak-Gajda, D., Bak, J.: Psychology of transport and road traffic safety. Difin, Warszawa (2010) (in Polish)
2. Khaisongkram, W., Raksincharoensak, P., Shimosaka, M., Mori, T., Sato, T., Nagai, M.: Automobile Driving Behavior Recognition Using Boosting Sequential Labeling Method for Adaptive Driver Assistance Systems. In: Dengel, A.R., Berns, K., Breuel, T.M., Bomarius, F., Roth-Berghofer, T.R. (eds.) KI 2008. LNCS (LNAI), vol. 5243, pp. 103–110. Springer, Heidelberg (2008)
3. Sakhi, S., Khettat, A., Hank, M., Elouardi, A., Bouaziz, S., Belhocine, M.: An Embedded Data Logger Based Vehicle Trajectory Monitoring System. In: International Conference on Multimedia Computing and Systems, Ouarzazate, Morocco, pp. 11–16 (2009)
4. Norma, N.H., Saad, Z., Isa, I.S., Ishak, S.: Software design and development for automotive online monitoring system. In: 7th IEEE International Colloquium on Signal Processing and its Applications, pp. 489–494 (2011)
5. Dong, Y., Hu, Z., Uchimura, K., Murayama, N.: Driver Inattention Monitoring System for Intelligent Vehicles: A Review. IEEE Transactions on Intelligent Transportation Systems, 596–614 (2011)
6. Aljaafreh, A., Khalel, M., Al-Fraheed, I., Almarahleh, K., Al-Shwaabkeh, R., Al-Etawi, S., Shaqareen, W.: Vehicular data acquisition system for fleet management automation. In: Proceedings of 2011 IEEE International Conference on Vehicular Electronics and Safety, pp. 130–133 (2011)
7. Yang, J., Chang, T.N., Hou, E.: Driver distraction detection for vehicular monitoring. In: IECON Proceedings (Industrial Electronics Conference), pp. 108–113 (2010)

8. Gavriilidis, A., Schwerdtfeger, T., Velten, J., Schauland, S., Höhmann, L., Haselhoff, A., Boschen, F., Kummert, A.: Multisensor data fusion for advanced driver assistance systems - The active safety car project. In: 7th International Workshop on Multidimensional (nD) Systems (2011)
9. Nabi, Z., Alvi, A., Mehmood, R.: Towards Standardization of In-Car Sensors. In: Strang, T., Festag, A., Vinel, A., Mehmood, R., Rico Garcia, C., Röckl, M. (eds.) Nets4Trains/Nets4Cars 2011. LNCS, vol. 6596, pp. 216–223. Springer, Heidelberg (2011)
10. Scialfa, C.T., Guzy, L.T., Leibowitz, H.W., Garvey, P.M., Tyrrell, R.A.: Age differences in estimating vehicle velocity. Psychology and Aging 6, 60–66 (1991); Sommer, M., Herle, M., Häusler, J., Risser, R., Schützhofer, B., Chaloupka, C.: Cognitive and personality determinants of fitness to drive. Transportation Research Part F: Traffic Psychology and Behaviour 11, 362–375 (2008)
11. Veitch, J.A.: Light source flicker: When and how does it affect us? In: Conference Voices for SSL Efficiency 2010 DOE Solid-State Lighting Market Introduction Workshop, Philadelphia (2010)
12. Task Force of the European Society of Cardiology the North American Society of Pacing Electrophysiology, "Standards of Measurement, Physiological Interpretation, and Clinical Use". Circulation 93, 1043–1065 (1996), doi:10.1161/01.CIR.93.5.1043
13. Filary, S.: Security in public transport. Wydawnictwo Wyższej Szkoły Bezpieczeństwa, Poznań (2010) (in Polish)
14. Mitas, A.W., Czapla, Z., Bugdol, M., Ryguła, A.: Recording and assessing biometric parameters of drivers to improve the road safety. Seria Transport, vol. 66, pp. 72–77. Zeszyty Naukowe Politechniki Ślaskiej, Gliwice (2010) (in Polish)
15. Piotrowski, Z.: Robust Algorithm for Heart Rate (HR) detection and Heart Rate Variability (HRV) Estimation. Acta Physica Polonica A, Optical and Acoustical Methods in Science and Technology 118(1), 131–135 (2010)
16. Wiciak, J., Borkowski, B.: The concept of a system for identification of the dangerous areas for people with visual dysfunction. Acta Physica Polonica A 119(6A), 921–924 (2011)
17. Iwaniec, M., Wiciak, J.: Prediction of vibroacoustical parameters on a cruise ship. Acta Acustica united with Acustica 88(5), 722–725 (2002)
18. Pankowska, M., Hojan-Jezierska, D., Skrodzka, E., Szymiec, E., Kubisz, L., Swidzinski, T., Wicherc, A.: Effect of ELF magnetic stimulation on distortion product of otoacoustic emission in tinnitus patients. Acta Physica Polonica A 119(6A), 1035–1039 (2011)

Toward Revocation Data Handling Efficiency in VANETs*

Carlos Gañán, Jose L. Muñoz, Oscar Esparza, Jorge Mata-Díaz, and Juanjo Alins

Universitat Politécnica de Catalunya, Departament Enginyeria Telemática
{carlos.ganan,jose.munoz,oesparza,jmata,juanjo}@entel.upc.es

Abstract. Vehicular Ad Hoc Networks (VANETs) require some mechanism to authenticate messages, identify valid vehicles, and remove misbehaving ones. A Public Key Infrastructure (PKI) can provide this functionality using digital certificates, but needs an efficient mechanism to revoked misbehaving/compromised vehicles. The IEEE 1609.2 standard states that VANETs will rely on the use of certificate revocation lists (CRLs) to achieve revocation. However, despite their simplicity, CRLs present two major disadvantages that are highlighted in a vehicular network: CRL size and CRL request implosion. In this paper, we point out the problems when using CRLs in this type of networks. To palliate these issues, we propose the use of Authenticated Data Structures (ADS) that allow distributing efficiently revocation data. By using ADS, network entities can check the status of a certificate decreasing the peak bandwidth load in the distribution points.

Keywords: Certification, PKI, Authenticated Data Structures.

1 Introduction

In the last decade, wireless communication between vehicles have drawn extensive attention for their promise to contribute to a safer, more efficient, and more comfortable driving experience in the foreseeable future. This type of communications have stimulated the emergence of Vehicular ad hoc networks (VANETs) which consist of mobile nodes capable of communicating with each other (i.e. Vehicle to Vehicle Communication -V2V communication) and with the static infrastructure (i.e. Vehicle to Infrastructure Communication -V2I communication). To make these communications feasible, vehicles are equipped with on-board units (OBUs) and fixed communication units (road-side units, RSUs) are placed along the road. Applying short range wireless technology based on IEEE 802.11, multi-hop communication facilitates information exchange among network nodes that are not in direct communication range [1].

* This work is funded by the Spanish Ministry of Science and Education under the projects CONSOLIDER-ARES (CSD2007-00004) and TEC2011-26452 "SERVET", and by the Government of Catalonia under grant 2009 SGR 1362.

A. Vinel et al. (Eds.): Nets4Cars/Nets4Trains 2012, LNCS 7266, pp. 80–90, 2012.

However, the open-medium nature of these networks and the high-speed mobility of a large number of vehicles make necessary the integration of primary security requirements such as authentication, message integrity, non-repudiation, and privacy [2]. Without security, all users would be potentially vulnerable to the misbehavior of the services provided by the VANET. Hence, it is necessary to evict compromised, defective, and illegitimate nodes. The basic solution envisioned to achieve these requirements is to use digital certificates linked to a user by a trusted third party. These certificates can then be used to sign information. Most of the existing solutions manage these certificates by means of a central Certification Authority (CA) [3]. According to IEEE 1609.2 standard [4], vehicular networks will rely on the public key infrastructure (PKI). In PKI, a CA issues an authentic digital certificate for each node in the network. Therefore, an efficient certificate management is crucial for the robust and reliable operation of any PKI. A critical part of any certificate-management scheme is the revocation of certificates.

Regarding the revocation of these certificates, some proposals allow revocation without the intervention of the infrastructure at the expense of trusting other vehicles criteria; and other proposals are based on the existence of a central entity, such as the CA, which is in charge of taking the revocation decision for a certain vehicle. Again, according to the IEEE 1609.2 standard [4], vehicular networks will rely on the existence of a CA. In this sense, it is stated that these networks will depend on certificate revocation lists (CRLs) and short-lived certificates to achieve revocation. CRLs can be seen as black lists that enumerate revoked certificates along with the date of revocation and, optionally, the reasons for revocation.

As the network scale of VANETs is expected to be very large and to protect the privacy of users each vehicle has many temporary certificates (or called pseudonyms), the CRLs are expected to be quite large. Moreover, CRLs have also associated a problem of request implosion, i.e., vehicles may become synchronized around CRL publication instant, as they may request CRL at or near the moment of publication. This burst of requests may cause network congestion that may introduce longer latency in the process of validating a certificate. To reduce the potential network and computational overhead imposed by any CRL distribution mechanism, some optimizations for organizing, storing, and exchanging CRL information have been proposed. In [2,5], it is proposed a way of compress CRLs using Bloom filters. Their method reduces the size of a CRL by using about half the number of bytes to specify the certificate serial number for revocation. However, the use of this probabilistic structure has associated a false positive rate that diminishes the efficiency of the revocation service.

In this paper, we explore the benefits of using authenticated data structures (ADS), such as binary trees or skip lists, to manage revocation data in VANETS. These structures are a model of computation where untrusted responders answer certificate status queries on behalf of the CA and provide a proof of the validity of the answer to the user. Although VANETs can greatly benefit from the use of ADSs, to the best of our knowledge there has been no proposal of deploying the revocation service by means of an ADS. By using these structures, both CRL issues are palliated: the CA is no longer a bottleneck as there are several responders that act on its behalf; and the revocation data can be checked without downloading the whole CRL.

2 CRLs' Problematic in VANETs

As stated in the trial-use standard [4], for a certificate authority (CA) to invalidate a vehicle's certificates, the CA includes the certificate serial number in the CRL. The CA then distributes the CRL so that vehicles can identify and distrust the newly revoked vehicle. The distribution should spread quickly to every vehicle in the system.

However, the distribution itself poses a great challenge due to the size of the CRL. As a CRL is a list containing the serial numbers of all certificates issued by a given certification authority (CA) that have been revoked and have not yet expired, its distribution causes network overhead. Moreover, the CRL size increases dramatically if only a small portion of the OBUs in the VANET is revoked. To have an idea of how big the CRL size can be, consider the case where 1% of the total number of the OBUs in the United States is revoked. Recall that in a VANET, each vehicle owes not only an identity certificate, but also several pseudonyms. The number of pseudonyms may vary depending on the degree of privacy and anonymity that it must be guaranteed. According to Raya, Papadimitratos, and Hubaux in [5] the OBU must store enough pseudonyms to change pseudonyms about every minute while driving. This equates to about 43,800 pseudonyms per year for an average of two hours of driving per day. In the United States alone, 255,917,664 "highway" registered vehicles were counted in 2008, of which 137,079,843 passenger cars [6]. In this case, the CRL would contain around 100 billion revoked certificates. Assuming that certificates can be identified by a 16 byte fingerprint (the size of one AES block), the CRL size would be of 1,7 TB approximately. Only the amount of memory necessary to storage this CRL makes it impossible its deployment. Therefore, the CRL size has to be reduced.

The CRL size can be reduced by using regional CAs. However, there appears a trade-off between the size of the CA region and size of the CRL, as well as the management complexity of the entire PKI system for VANETs. The least complicated region to manage would be a single large area, such as the entire United States, with a single CA responsible for every certificate and pseudonym. However, this gives place to CRLs of several terabytes. Therefore, it is necessary to divide the CRL information according to regional areas. In this sense, if we divide the entire United States by cities (i.e. 10,016 cities according to the U.S. census bureau), the CRL size is reduced to around 170 Mbytes. Using the 802.11a protocol to communicate with RSUs in range, vehicles could have between 10-30 Mbps depending on the vehicle's speed and the road congestion. Therefore, in the best case a vehicle will need more than 45 seconds to download the whole CRL. Under non-congested conditions, any vehicle should be able to contact the infrastructure for more than 45 seconds, and therefore download the CRL. In scenarios where vehicles are not able to keep a permanent link with the infrastructure for this amount of time, techniques such as Bloom filter or Digital Fountain Codes could be used to download the CRL. Therefore, though the problem of having a huge CRL is mitigated by the use of such techniques, the restraints imposed by the distribution affect the freshness of the revocation data.

A direct consequence of this significant time to download a CRL is that a new CRL cannot be issued very often, so its validity period has to be shortened. This validity period directly determines how often a vehicle has to update the revocation information. Therefore, the validity period of the CRL is critical to the bandwidth consumption. In this context, it appears another trade-off between the freshness of revocation information and the bandwidth consumed by downloading CRLs. Large validity periods will decrease the network overhead at expenses of having outdated revocation information. Small validity periods will increase the network overhead but users will have fresh information about revoked certificates. As CRLs cannot be issued every time there is a new revoked certificate, vehicles will be operating with revocation information that is not comprehensive. Therefore, they will be taking certain risk of trusting a certificate that could be potentially revoked.

3 Using Authenticated Data Structures for Certificate Revocation in VANETs

By replicating revocation data at untrusted responders near users, VANETs can enhance its performance but that replication causes a major security challenge. Namely, how can a vehicle verify that the revocation data replicated at the RSUs are the same as the original from the CA? A simple mechanism to achieve the authentication of replicated revocation data consists of having the digitally sign each revocation entry and replicating the CA signature too. However, in VANETs where the revocation data evolves rapidly over time, this solution is inefficient. To achieve higher communication and computation efficiency, we propose the use of authenticated data structures (ADS) to handle the revocation service in VANETs. ADSs are a model of computation where untrusted responders answer certificate status queries on behalf of the CA and provide a proof of the validity of the answer to the user. In this section, first we introduce the architecture necessary to adopt ADSs. Then, we describe different ADSs and their main benefits.

3.1 System Architecture

The system architecture to support ADSs consists in an adaptation of a PKI system to the vehicular environment. The ADS model involves a structured collection $\mathcal{R}$ of revoked certificates and three parties: the certification authority

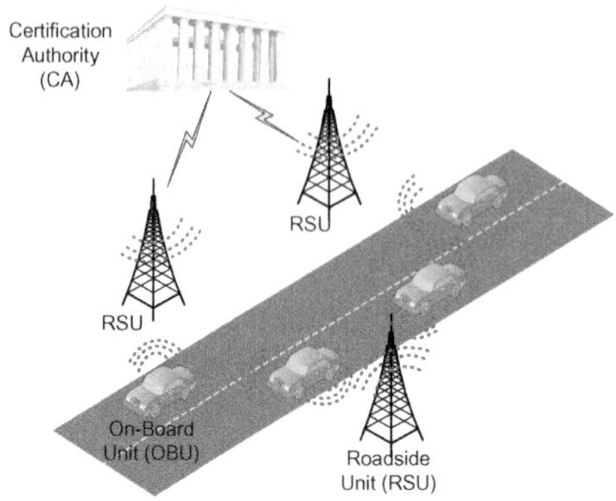

Fig. 1. System Architecture

(CA), the road side units (RSUs), and the vehicles. A repertory of query operations and optional update operations are assumed to be defined over $\mathcal{R}$. These three parties present a hierarchical architecture (see Fig. 1) which consists of three levels: the CA is located at level 1, as it is the top of the system. RSUs are located at level 2. Finally, the on-board units (OBUs) are located at the bottom of the hierarchy. Note that without loss of generality we consider a single the central trusted authority at the root, but it could be further divided into different state level trusted authorities and additionally a group of city level trusted authorities can be placed under every state authority.

The main tasks of each entity are:

1. The CA is responsible for generating the set of certificates that are stored in each OBU. It is also responsible for holding the original version of $\mathcal{R}$ and making it accessible to the rest of the entities. By definition of TTP, the CA should be considered fully trusted by all the network entities, so it should be assumed that it cannot be compromised by any attacker. In fact, in our proposal the CA is the only trusted entity within the network. Whenever an update is performed on $\mathcal{R}$, the CA produces structure authentication information, which consists of a signed time-stamped statement about the current version of $\mathcal{R}$.
2. RSUs are fixed entities that are fully controlled by the CA. They can access the CA anytime because they are located in the infrastructure-side, which does not suffer from disconnections. RSUs maintain a copy of $\mathcal{R}$. They interact with the CA by receiving from the CA the updates performed on $\mathcal{R}$ together with the associated structure authentication information. RSUs also interact with vehicles by answering queries on $\mathcal{R}$ posed by the vehicles. In addition to the answer to a query, RSUs also return answer authentication information, which consists of (i) the freshest structure authentication information issued by the CA; and (ii) a proof of the authenticity of the answer. If the CA considers that an RSU has been compromised, the CA can revoke it.

3. OBUs are in charge of storing all the certificates that a vehicle possesses. An OBU has abundant resources in computation and storage and allows any vehicle to communicate with the infrastructure and with any other vehicle in its neighborhood. OBUs pose queries on $\mathcal{R}$, but instead of contacting the CA directly, it contacts the RSU in range. However, OBUs only trust the CA and not the RSU about $\mathcal{R}$. Hence, it verifies the answer from the RSU using the associated answer authentication information.

3.2 System Requirements

- *Low computational cost*: The computations performed internally by each entity (CA, RSU, and OBU) should be simple and fast.
- *Low communication overhead*: CA-to-RSU communication (update authentication information) and RSU-to-OBU communication (answer authentication information) should be as small as possible.
- *High security*: the authenticity of the answers given by a RSU should be verifiable.

3.3 Authenticated Data Structures

Several ADSs have been proposed in the literature (mainly in the context of data base management) that fulfill the aforementioned requirements. In this section, we describe a repertoire of ADSs and to what extent they are capable of improving the revocation service.

Merkle Hash Trees. A Merkle hash tree (MHT) [7] is essentially a tree structure that is built with a collision-resistant hash function to produce a short cryptographic description of $\mathcal{R}$. The leaf nodes hold the hash values of the data of interest, i.e., the serial number of the revoked certificates (SN_1,SN_2,...,SN_n); and the internal nodes hold the hash values that result from applying the hash function to the concatenation of the hash values of its children nodes. In this way, a large number of separate data can be tied to a single hash value: the hash at the root node of the tree. MHTs can be used to provide an efficient and highly-scalable way to distribute revocation information.

A sample MHT is presented in Figure 2(a). The authentication of an element is performed using a verification path, which consists of the sibling nodes of the nodes on the path from the leaf associated with the element to the root of the tree. The root value is signed and the collision-resistant property of the hash function is used to propagate authentication from the root to the leaves. This construction is simple and efficient and achieves signature amortization, where only one digital signature is used for signing a large collection of data. The hash tree uses linear space and has $O(log\ n)$ (where n denotes the number of revoked certificates) proof size, query time and verification time. An ADS based on hash trees can also achieve $O(log\ n)$ update time.

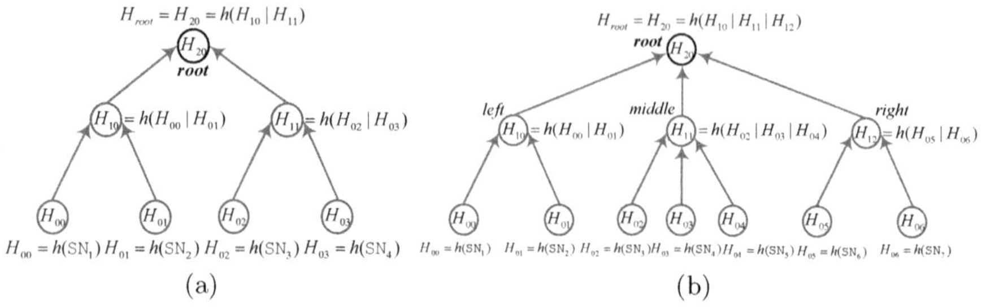

Fig. 2. Sample trees (a) MHT (b) 2-3

2-3 Trees. A standard 2-3 tree [8] is a tree where all leaves are at the same height and each node (except leaves) has two or three children. It has the nice property that leaf removal and insertion incur only logarithmic complexity because these operations only involve the nodes related to the path from the relevant leaf to the root.

Each leaf of such a 2-3 tree stores an element of set $\mathcal{R}$, and each internal node stores a one-way hash of its children's values. Thus, the CA-to-RSU communication is reduced to $O(1)$ entries, since the CA sends insert and remove instructions to the RSUs, together with a signed message consisting of a timestamp and the hash value of the root of the tree. RSUs respond to a membership query for an element SN_i as follows: if SN_i is in $\mathcal{R}$, then RSUs provide the path from the leaf storing SN_i to the root, together with all the siblings of the nodes on this path; else (SN_i is not in$\mathcal{R}$), RSUs provide the leaf-to-root paths from two consecutive leaves storing SN_j and SN_k such that $j < i < k$, together with all siblings of the nodes on these paths. By tracing these paths, OBUs can recompute the hash values of their nodes, ultimately recomputing the hash value for the root, which is then compared against the signed hash value of the root for authentication. As with MHTs, these trees achieve $O(log\ n)$ proof size, query time, update time and verification time.

One-Way Accumulator. One-way accumulator (OWA) functions [9] allow a CA to digitally sign a collection of objects as opposed to a single one. The main advantage of this approach is that the validation of a response takes constant time and requires computations simple enough to be performed in resource-constrained devices. This type of ADS achieves a tradeoff between the cost of updates at the CA and queries at the RSUs, with updates taking $O(k + log(\frac{n}{k}))$ time and queries taking $O(\frac{n}{k})$ time, for any fixed integer parameter $1 \leq k \leq n$. For instance, one can achieve $O(\sqrt{n})$ time for both updates and queries.

Skip Lists. Skip lists [10] are probabilistic ADSs that provide an alternative to balanced tree. Skip lists are sorted linked lists with extra links, designed to allow fast search in $\mathcal{R}$ by taking "shortcuts". The main idea is to enhance linked lists, which connect each element in the data sequence to its successor, by also connecting some elements to successors further down the sequence. Roughly

Table 1. Comparison of the main ADS vs traditional revocation mechanisms

method	space	update time	update size	query time	query size	verifying time
CRL	$O(n)$	$O(1)$	$O(1)$	$O(n)$	$O(n)$	$O(n)$
OCSP	$O(n)$	$O(n)$	$O(n)$	$O(1)$	$O(1)$	$O(1)$
MHT	$O(n)$	$O(log\ n)$	$O(1)$	$O(log\ n)$	$O(log\ n)$	$O(log\ n)$
2-3 tree	$O(n)$	$O(log\ n)$	$O(1)$	$O(log\ n)$	$O(log\ n)$	$O(log\ n)$
Skip Lists	$O(n)$	$O(log\ n)$	$O(1)$	$O(log\ n)$	$O(log\ n)$	$O(log\ n)$
OWA	$O(n)$	$O(k + log(\frac{n}{k}))$	$O(k)$	$O(\frac{n}{k})$	$O(1)$	$O(1)$

half of the elements have links to their two-hop successor, roughly a quarter of the elements have links to their four-hop successor, and so on. As a result, during traversal from SN_i to element SN_j, the traversal path follows repeatedly the longest available link from the current element that does not overshoot the destination SN_j, and thereby reaches SN_j in fewer steps than would be possible by just traversing every intervening element between SN_i and SN_j. Compared with balanced trees, a skip list presents the following benefits:

- It is easy to implement and practically efficient in search, especially update time.
- It is space compact, where space is allocated when needed, while empty space is preserved in balanced tree.
- It is main memory index, while balanced tree are disk-based index.

Finally, Table 1 shows a comparison of the asymptotic performance of the main ADS versus traditional revocation mechanisms such as CRL or OCSP. Note that with ADSs, the revocation service can be greatly improved both in computation and communication overhead.

3.4 Certificate Status Validation Protocol

The certificate status validation protocol consists in three stages.

1. *Revocation Service Setup*: The CA creates a CRL by appending the serial number of any revoked certificate. Then, it computes the corresponding ADS from the set $\mathcal{R}$ of revoked certificates contained in the CRL. Once the ADS is computed, the CA signs the resulting time-stamped digest of the data structure, i.e., a collision resistant succinct representation of the data structure. The digest is transmitted to all the RSUs via a secure wireline together with the corresponding CRL. RSUs can either implement a push or pull protocol to transmit the digest to the vehicles in range.
2. *Certificate Status Updating*: Depending on the CA's policy, when an update is necessary, the CA recomputes the ADS and generates a new signed digest that is transmitted to the RSUs. Note that depending on the ADS, the data structure should be computed again or only update and delete operations should be performed. The new ADS is transmitted to the RSUs again, so that they could answer to validation queries.

3. *Certificate Status Querying*: OBUs query any RSU in range about the status of a particular certificate (SN_i). If $SN_i \in \mathcal{R}$, then the RSU computes the path necessary to allow OBUs to compute the digest and check that it matches the signed digest. If $SN_i \notin \mathcal{R}$, then the RSU computes the path of two consecutive certificates in $\mathcal{R}$ and transmit them to the requesting OBU. This OBU can then recompute the digest for both revoked certificates and be sure that $SN_i \notin \mathcal{R}$.

4 Evaluation

In the following, we compare the communication costs of using ADSs with the tradition CRL mechanism. To that end we define a set of parameters (see Table 2).

Table 2. Notation

Parameter	Meaning of the parameter
N	Total number of certificates ($n = 3,000,000$)
k	Average number of certificates handled by a CA ($k = 30,000$)
p	Percentage of revoked certificates($p = 0.1$)
q	Number of certificate status queries issued per day ($q = 3,000,000$)
T	Number of updates per day ($T = 1$)
s_{SN}	Size of a serial number ($s_{SN} = 20$)
s_{sig}	Size of a signature ($s_{sig} = 1,000$)
s_{hash}	Size of the hash function ($s_{hash} = 128$).

Using this notation, the CRL daily update cost is $T \cdot n \cdot p \cdot s_{SN}$ as each CA sends the whole CRL to the corresponding RSUs in each update. The CRL daily query cost is $q \cdot p \cdot k \cdot s_{SN}$ as for every query the RSU sends the whole CRL to the querying OBU. When using ADS, these costs are drastically reduced. Note that no matter the type of ADS, OBUs do not have to download the whole CRL, and they only download status information about the certificate they want to operate with. Regarding MHTs, the RSUs have to recompute the tree in each update, so that daily update cost is $T \cdot n \cdot p \cdot s_{SN}$. However, to answer an OBU's query the RSU only needs to send up to $1 + log_2(pk)$ numbers, resulting in $q \cdot s_{hash}(1 + log_2(pk))$ bits. In the case of 2-3 trees, to update the directory, the CA sends difference lists of total daily length of $\frac{n \cdot p \cdot s_{SN}}{365} + T \cdot s_{sig}$; and answer to OBUs' queries results in $2 \cdot q \cdot s_{hash} \cdot log_2(pk)$ bits. Similarly, skip lists need $2log_2\lceil pk \rceil$ number to answer an OBU's query and the same update cost than the 2-3 tree. With OWAs, the size of answer are drastically reduced to roughly s_{sig}, and the update cost depends on the accumulator configuration. We use Matlab R2011b to evaluate these costs.

Note that the costs will vary mainly depending on the total number of revoked certificates, the update rate and the number of queries. Figure 3(a) shows how the CA-to-RSU update communication costs of the different revocation mechanisms

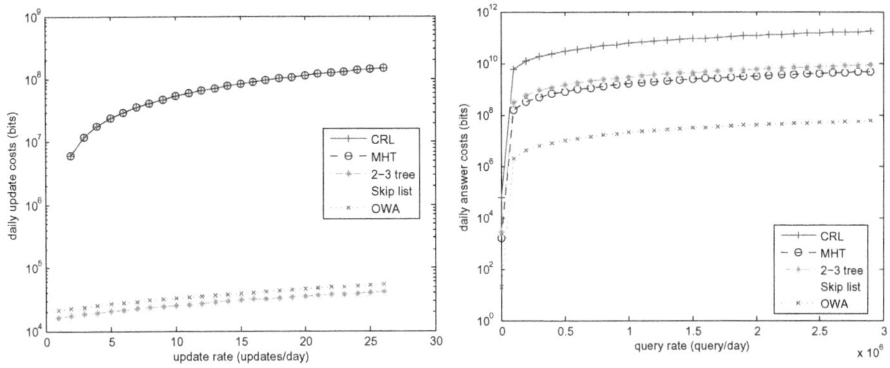

(a) (b)

Fig. 3. (a) Daily CA-to-RSU update costs vs. update rate, (b) RSU-to-OBU query cost vs query rate

depend on the update rate (all other parameters are held constant). Note that any ADS is much more robust and efficient than CRL, even allowing once per hour updates. Regarding the query costs, as ADSs have smaller proof to validate the status of a certificate they provide a more bandwidth efficient solution than CRL (see Fig. 3(b)).

5 Conclusions

In this paper, we consider the problem of certificate authentication and revocation in VANETs. We have proposed the use of authenticated data structures to handle the revocation service over VANETs. After discussing the issues of deploying CRLs in these environments, we show that ADSs are more robust to changes in parameters, and allow higher update/query rates than traditional revocation mechanisms. In addition, the adoption ADS reduces both the communication and the computational overhead in the OBUs. For our future work, we will investigate the use of mobile repositories under the context of the proposed schemes.

References

1. Jiang, D., Delgrossi, L.: IEEE 802.11p: Towards an International Standard for Wireless Access in Vehicular Environments. In: IEEE Vehicular Technology Conference, VTC Spring 2008, pp. 2036–2040 (May 2008)
2. Raya, M., Hubaux, J.-P.: The security of vehicular ad hoc networks. In: Proceedings of the 3rd ACM Workshop on Security of Ad Hoc And Sensor Networks, SASN 2005, pp. 11–21 (2005)
3. Papadimitratos, P., Buttyan, L., Hubaux, J.-P., Kargl, F., Kung, A., Raya, M.: Architecture for secure and private vehicular communications. In: 7th International Conference on ITS Telecommunications, pp. 1–6 (June 2007)

4. IEEE trial-use standard for wireless access in vehicular environments - security services for applications and management messages. IEEE Std 1609.2-2006, pp. 1–105 (2006)
5. Haas, J.J., Hu, Y.-C., Laberteaux, K.P.: Design and analysis of a lightweight certificate revocation mechanism for vanet. In: Proceedings of the Sixth ACM International Workshop on VehiculAr InterNETworking, VANET 2009, pp. 89–98. ACM, New York (2009)
6. Bureau of Transportation Statistics U.S. Department of Transportation. Number of U.S. aircraft, vehicles, vessels, and other conveyances (2009), http://www.bts.gov/publications/national_transportation_statistics/html/table_01_11.html (Online; accessed July 31, 2011)
7. Merkle, R.C.: A Certified Digital Signature. In: Brassard, G. (ed.) CRYPTO 1989. LNCS, vol. 435, pp. 218–238. Springer, Heidelberg (1990)
8. Naor, M., Nissim, K.: Certificate Revocation and Certificate Update. IEEE Journal on Selected Areas in Communications 18(4), 561–570 (2000)
9. Benaloh, J.C., de Mare, M.: One-Way Accumulators: A Decentralized Alternative to Digital Signatures. In: Helleseth, T. (ed.) EUROCRYPT 1993. LNCS, vol. 765, pp. 274–285. Springer, Heidelberg (1994)
10. Pugh, W.: Skip lists: a probabilistic alternative to balanced trees. Commun. ACM 33, 668–676 (1990)

Cluster Based Data Aggregation in Vehicular Adhoc Network

Muhammad Shoaib[1], Wang-Cheol Song[1,*], and Keun Hyung Kim[2]

[1]Department of Computer Engineerng
[2] Department of Mangement Information Systems
Jeju Natinal University
Jeju, 690-756 Republic of Korea
muhammad.shoaib@live.com, {philo,khkim}@jejunu.ac.kr

Abstract. Information aggregation is used to fuse interrelated data items collected from different nodes before redistributing them. Thus, using aggregation the number of transmissions and the communication overhead can be reduced significantly. Particularly for applications which require periodic dissemination of information into a large region, aggregation becomes prerequisite. Let us consider an example of self-organized infrastructure less traffic information system where each vehicle periodically disseminates its speed, position and information about road conditions. The aim is to design such data aggregation technique that not only use low bandwidth for data dissemination after aggregation process but also use low bandwidth for data collection from source nodes as well. As VANET environment has limited bandwidth this data aggregation becomes the need of hour for efficient data processing and dissemination without network overhead. In this work we have proposed a cluster based data aggregation technique where vehicles are divided into autonomous clusters headed by cluster-head. Data is send to cluster-head that perform the aggregation of data and dissimulate it into network.

Keywords: Data Aggreigation in VANET; VANET applicaitons; VANET data dessiminaiton; cluster based data aggregation; Autonomus Clustering.

1 Introduction

Inter-vehicle communication (IVC) is an enabling technology for many applications that is being used actively in safety, enhanced driving, or entertainment applications. Data Aggregation Techniques for VANETs need a careful design for being able to cope with the variety of requirements emerging from the highly dynamic network structure and the applications themselves. One common problem in IVC constitutes the limited bandwidth of the wireless medium. In case of high vehicle density many cars compete for the wireless medium, thus exceeding the available bandwidth if every car disseminates a large amount of application level data. Moreover, many

* Corresponding author.

A. Vinel et al. (Eds.): Nets4Cars/Nets4Trains 2012, LNCS 7266, pp. 91–102, 2012.

applications need to flood data into a large geographic region which amplifies the problem of the limited bandwidth [1].

Information aggregation is one possibility to address this problem. The idea behind aggregation is that many applications do not need atomic information, such as exact position and speed values from each vehicle. Instead, it is sufficient to have an abstract, aggregated view of the sensed phenomenon. The level of the abstraction is application dependent and influences the amount of bandwidth saved. Consider for example a traffic information system which is used to inform drivers about road conditions several kilometers ahead. The native way to implement such a system is to disseminate the own position periodically, together with all other atomic information received from other vehicles. It is easy to see that such a system is not scalable. In addition, a driver is not interested in such atomic information, but wants to know the average speed on the road, and where a possible traffic jam is exactly located to be able to react accordingly.

Such a subsumption of similar atomic values in one aggregated value (e.g., average speed) is called data fusion. For the effectiveness of aggregation the spatial and temporal correlation of data is of key importance. In traffic jams, the velocity of the vehicles is almost identical. Thus, all atomic values in the whole traffic jam region can be aggregated to one average, without significant loss of accuracy. This way, only a fraction of information has to be disseminated, thereby significantly reducing communication overhead.

As we have seen, the correlation of data is important for aggregation and the correlation factor and its dimension depends on actual traffic and road conditions. Therefore, the key idea of this work is a completely infrastructure-free aggregation mechanism, which enables to aggregate data purely based on their correlation within the network without using any external sources. This way the proposed protocol achieves a very high accuracy in terms of reducing the size of information that needs to be dissimilated especially broadcasted or geo-casted in the network. Other proposed aggregation approaches for VANETs rely on some kind of fixed structure (such as road segments) in which data is aggregated, regardless of their correlation. Thus, such approaches always have to handle with a trade-off between accuracy and communication overhead: with small segment sizes a better approximation is possible, but more information has to be communicated. Having large segments, the deviation of the aggregated values from the actual situation is higher, resulting in reduced accuracy.

This paper is meant as a step towards a deeper understanding of these fundamental issues. Previous work has considered individual, specific aggregation schemes and dissemination mechanisms, and has evaluated them in specific situations and environments, using simulations and experiments. Here, we consider the zone based clustering environment where data aggregation and dissemination is done by cluster-head instead of all vehicular nodes personally. Our primary focus is on reducing the size of information that is need to be disseminate to ensure that all information can be broadcasted in lower bandwidth as well in any kind of VANET environment. For data aggregation we proposed a weighted geomantic mean algorithm where we define the weights of the values as the function of time. Instead of the other approaches that uses different forms of anathematic means explained in [2, 3, 4] we used geomantic mean to ensure that outlier values don't effect on aggregation technique.

To fulfill our aim we use the cluster based technique where we first don't encourage data aggregation and disseminate by each node but we want to delegate this responsibility to cluster-head. In our experiments we show that our approach reduces the data size that eventually reduces the network traffic on a network. Our result provides value able hints in development and usage of cluster oriented technique in conjunction with data aggregation technique.

The rest of the paper organized as follows; in section 2 we present related work in data aggregation field for vehicular network. In section 3 we explain our functional equation based mechanism for data aggregation followed by cluster formulation for vehicular network in section 4. In section 5 we present our experimental details and obtained results followed by conclusion in section 6 of the paper.

2 Related Work

Aggregation in Wireless Sensor Network (WSNs) is a well-researched area, but WSNs also differ notably from VANETs both regarding network properties and requirements, which renders simple transfer of aggregation schemes infeasible. First of all, in most sensor network aggregation schemes it is assumed that there is only one base station which initializes a query to gather aggregated data sensed by nodes. Afterward, all aggregated data is propagated to this base station as the data sink. This characteristic is exploited by tree-based aggregation where the root node of the tree is the data sink. In contrast to this, in vehicular networks all nodes are data sinks. Therefore, such aggregation schemes are not appropriate for vehicle-to-vehicle communication. Furthermore, Node based data aggregation techniques needs a significant amount of communication overhead for transferring each and every record to all vehicle via broadcasting. This overhead increases with growing node mobility. Therefore, such static topology hierarchies cannot be used in vehicular networks, where we face very high node velocities and network dynamics.

In [6], Author discussed about relationship and data dissemination and data aggregation they proposed a model for balancing data amount with respect to network band width. The paper studies the direct impact of aggregation model on the performance of network and data dissimilation. Instead of performing exponents author had adopted the mathematical way to prove their theory. According to the authors, the level or aggregation is directly proportional to network bandwidth.

In [8] author proposed a time-efficient local data aggregation algorithm for ubiquities sensor networks that aggregates delay-constrained data within a given time deadline in clustered wireless sensor networks. Approach consists of two phases. At the first step author had design a zone-based fast data aggregation tree (ZFDAT) to eliminate unnecessary packets being forwarded too many receiver nodes and avoid long detour paths until cluster-head, where cluster-head is local control centre to coordinate the data transmissions in the cluster. Next, In the second step optimal link scheduling algorithm to minimize aggregation time by given variable length of time slots for all links in the ZFDAT has been explained the core objective of the approach was to minimize the aggregation time under energy constraint and outperforms the similar existing algorithms. In particular, this approach distributes the total aggregation time, and then it finally also reduces the heavy processing load at the cluster-head.

Another approach is explained in [3] for the decentralized discovery of free parking places. Vehicles use periodic broadcasts (beacons) to disseminate information about free parking slots. In that work, aggregation is performed over a hierarchical quad-tree. Consequently, the map is divided into no overlapping areas and information is organized in different levels. This way the communication overhead can be reduced drastically because aggregated information is disseminated into remote regions with increasing aggregation level (i.e., with decreasing accuracy). However building a tree based approach required pre-defined map information and it can't work in dense area like on roads or traffic signals where data needs to be disseminate to the other vehicles. In such situations on the one side vehicles are in large amount and on the other side data size also keep increasing that consequently overflow from provided bandwidth. Lochert et al. in [4] take a similar hierarchical approach on aggregating free parking slots using globally known map data for segmentation. One major advantage of their system is the usage of an adapted version of Flajolet-Martin sketches to achieve a probabilistic but duplicate insensitive sum of free parking spaces [3]. Aggregates can therefore be arbitrarily combined and re-combined without counting free parking slots multiple times.

Another approach is explained in [5] for the decision making of data aggregation in vehicular environment. Vehicles use fuzzy reasoning as a holistic yet intuitive approach to take the aggregation decision. Fuzzy reasoning allows a flexible aggregation decision that may consider all values contained in the aggregates for an aggregation decision. As the result of fuzzy function the output is presented in HIGH, LOW and MEDIUM aggregation. The parameters that are used for the fuzzy function in order to make the decision are road segment density and vehicle's speed. Dis-advantage of this approach is it works on whole network and doesn't create any sub-graphs of the network hence the data aggregation is done by each node and band-width problem remains the same as explained in the above stated work.

3 Functional Equation

Functional equations are equations where the unknowns are functions. A well-known example of functional equation is the following Cauchy equation:

$$\emptyset\,(x + y) = \emptyset(x) + \emptyset(y)$$

A function $\emptyset$ is a solution of this equation if, for any two values x and y, the application of φ to x + y equals the addition of the application of $\emptyset$ to x and to y. Therefore, the equation establishes conditions that functions φ have to satisfy. Typical solutions of this Cauchy equation are the functions $\emptyset(x) = \alpha x$ for an arbitrary value for α.

- In information fusion, functional equations can be used in two different contexts. Functional equations can be used when we need to define an aggregation operator and we know which basic properties it has to satisfy. We can express the conditions of such an operator using functional equations. The operator is then derived from the equations

- Functional equations can be used to study the properties of information fusion methods. This is so because they can characterize the operators. Here, a characterization consists of finding a minimum set of properties (a minimum set of equations) that uniquely implies the operator. It is important to say that the set of properties that imply an operator is usually not unique.

In this section, we present an example of the application of functional equations to construct an operator that aggregates numerical information. We start by formalizing the problem, and then turn to the development of its solution. The example illustrates the main procedures to prove theorems by means of functional equations.

In a vehicular ad-hoc network, *m* different cars have been recording their *N* different sensors parameters to publish for their respective cluster head. Then, a cluster head has to aggregate the information of all the cars and broadcast final aggregated values of the obtained values. The first problem is how to formalize the aggregation process. We start by formalizing the available data and then the requirements for the aggregation process.

Available data is modeled as follows: x_j^i stands for the evaluated value that the *i*th vehicle recorded for the *j*th parameter. X stands for the whole matrix with the opinions of all the vehicles (X = $\{x_j^i\}$) and xi corresponds to the vector with the values of the *i*th parameter recorded by all vehicles. Similarly, x_j is the vector with all the vehicles to the *j*th parameters values. Table 3.1 illustrates the data supplied by the vehicles. In this table, the *i*th vehicle is denoted by V_i.

	$Perm_1$	$Perm_2$	......	$Perm_i$	.....	$Perm_N$
Veh_1	x_1^1	x_2^1	......	x_i^1	......	x_N^1
Veh_2	x_1^2	x_2^2	......	x_i^2	......	x_M^2
Veh_3	x_1^3	x_2^3	......	x_i^3	......	x_N^3
....	......		......		......	
Veh_j	x_1^j	x_2^j	......	x_i^j	......	x_N^j
	......		......	......	......	......
Veh_M	x_1^M	x_2^M	......	x_i^M	......	x_N^M
DM	$\mathbb{C}_1(x_1)$	$\mathbb{C}_2(x_2)$	......	$\mathbb{C}_i(x_i)$	......	$\mathbb{C}_N(x_N)$

Before we introduce aggregation functions $\{f_1(x_1), f_2(x_2), \dots\dots\dots f\ (x_M)\}$, we introduce weight. The typical operator is the weighted mean. In this operator, weights are used to represent importance of the information sources, time of Information Recording and life of Information. When a value is time bound and recorded for certain time period at a particular time *t* we have weighted aggregation operators. We characterize these operators below.

Definition 1: An operator is sensitive if, for $all\ k = 1,\dots,N, when\ a_k \neq \bar{a}_k$,

$$\mathbb{C}(a_1, a_2, a_3, a_3, \dots a_{k-1}, a_k, a_{k+1} \dots a_N) \neq \mathbb{C}(a_1, a_2, a_3, a_3, \dots a_{k-1}, \bar{a}_k, a_{k+1} \dots a_N)$$

That is, changes in one parameter imply changes in the outcome. This property implies the cancellativity of ◦, a weighted parameter.

Bajraktarevic's means are a family of operators that generalize weighted means. Their definition has resemblances with a quasi-arithmetic and geomantic mean with weights (the quasi-weighted mean or geomantic weighted mean) While in the weighted mean, g function, the weight $g_i(a_i)$ is constant for all a_i, in a Bajraktarevic's mean the weight is a function of the a_i. Such a function is expressed by $g_i(a_i)$in the next definition

Given functions g and W (where W is monotone increasing with inverse W −1 and g nonnegative), the Bajraktarevic's mean is defined as follows:

Given functions f and g such as g is monotone increasing with inverse g^{-1} and f nonnegative, the Bajraktarevic's mean is defined as follows:

$$f_i(a_i) = \{x_1, x_1, x_1, \dots \dots \dots . x_M\} \quad \forall\, i \in [1, M]$$

$$\mathbb{C}\,(a_1, \dots \dots, a_N) = g^{-1}\left(\frac{\sum_{i=1}^{N} f_i(a_i)\, g_i(a_i)}{\sum_{i=1}^{N} f_i(a_i)}\right)$$

$$\mathbb{C}\,(a_1, \dots \dots, a_N) = g^{-1}\sqrt[N]{\left(\frac{\prod_{i=1}^{N} f_i(a_i)^{g_i(a_i)}}{\prod_{i=1}^{N} g_i(a_i)}\right)}$$

The Bajraktarevic's mean becomes the quasi-weighted mean when $g_i(a_i)$ is a constant that solely depends on $(g_i(a_i) = t)$, and the arithmetic and geomantic mean when $g_i(a_i) = k$ for all i. Families of arithmetic-weighted means and geomantic-weighted means can be generated applying the functions in following Equation

$$G.M. = \sqrt[N]{\left(\prod_{i=1}^{N} f_i\,(a_i)\right)} = \sqrt[N]{f_1(a_1) f_2(a_2) f_3(a_3) \dots \dots \dots \dots \dots . f_N(a_N)}$$

Our next step is to define weightages for the values with respect to time and distance between source node and cluster head.

Definition 2: *Let d is the distance between source node and destination node and t is time difference between period of information being recorded and processed. We define raw weight as follow*

$$W = 1 - \left(\frac{1}{t \times d}\right)$$

And raw weight for scaling as follows;

$$RW = t \times d$$

We define set of raw weights for all values $\{x_1, x_2, x_3 \dots\dots\dots x_N\}$ *as* $\{rw_1, rw_2, rw_3 \dots\dots\dots rw_N\}$ *and* $m = max(rw)$

To scale values from {0, 1} we use the following formula

$$for\ a\ value\ x_i\ the\ weight\ w_i = \frac{\mathrm{m}}{rw_i} = \frac{\max(rw)}{t_i \times d_i}$$

Once the weights have been calculated for all values provided in the table 1, the decision maker (denoted by DM in the last row of Table 3.1) has to make a final assignment. This final assignment is expressed as a function $\mathbb{C}$ of X. That is, $\mathbb{C}(X)$ denotes all the assignments and $\mathbb{C}_i(X)$ corresponds to the final assignment to a particular parameter *j* therefore $\mathbb{C}(X) = (\mathbb{C}_1(X), \dots, \mathbb{C}_N(X))$.

4 Autonomous Clustering Scheme

A cluster is a subset of nodes and consists of one cluster-head and two or more cluster-members. Node ID of the cluster-head becomes cluster ID of the cluster. A cluster-head and other cluster-members should connect with each other with the same cluster ID. The cluster is managed by the number of members. The number of nodes in each cluster is always adjusted between the lower bound η and the upper bound λ which are set beforehand, respectively. Moreover, a node which is adjacent to other node in a different cluster ID is called a gateway node. A gateway node performs the communication and the information gathering with the neighbor cluster.

We show autonomous clustering scheme in Figure 1. In the figure, a circle denotes a node, and a part enclosed by a solid line represents the cluster. The cluster-head-based tree is constructed in the cluster, and each cluster is connected with the neighbouring cluster through the gateway nodes. Before explaining the cluster architecture we assume the following properties about the vehicular network:

- The vehicles in the network are moving vehicle that moves with respect to some speed *s* and towards some direction *d.*
- Vehicles have similar capabilities for sensing, processing and communication and slandered communication range of 180m.
- Nodes are assumed to know their location information via GPS-like devices or other positioning technique.

In Autonomous clustering scheme, the nodes organize themselves into local clusters which one node acts as the cluster head. Each no cluster head node (cluster member) joins the nearest cluster head and transmits its data to the cluster head. Cluster head nodes aggregate the reported data from cluster members and transmit the data to the sink node. We assume that proper transmission control schemes for medium access are available to minimize the effect of dynamic wireless channel.

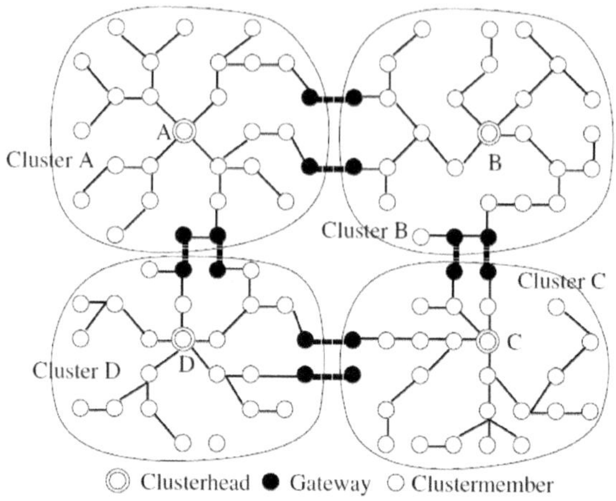

Fig. 1. Cluster formation

The operation of autonomous clustering scheme is divided into a series of consecutive tasks. Each task begins with a cluster setup phase that is initiated by the sink node. Figure 1 shows the state transitions of nodes in autonomous clustering scheme. After being connected to the network, vehicular nodes are in idle states. When a new membership request packet is broadcasted by a cluster-head, all nodes who receive the membership request came into ready state to evaluate the cluster coverage to become part of a cluster and got connected with its cluster head. The nodes that receive the broadcast message evaluate their willingness to become a part of cluster and set their willingness value for cluster head declaration, by using the cluster coverage and remaining energy.

We define μ the expected willingness of a vehicle v, as a function of distance and connection strength that is measured in form of hops to join a cluster or not.

Definition 3: *Let d is the distance between the cluster-head and cluster node and h is no of hops between cluster node and cluster-head. We define the willingness μ as follows;*

$$\mu = 1 - \left(\frac{1}{d \times h}\right)$$

Definition 4: *Let Ω is the function for selecting a cluster-head from a set of cluster-heads we define Ω as follow;*

$$\Omega(\mu_1, \mu_2 \ldots \ldots \mu_N) = \max\ (\mu_1, \mu_2 \ldots \ldots \mu_N)$$

When the threshold time expires and node doesn't receive any broadcast packet from any cluster head, the node then it-self broadcast the request to its neighbours to become part of some cluster that has the cluster head. The nodes that receive the broadcast message from their neighbourhood respond back with its cluster-head by providing the number of hops and distance between cluster-head and itself along with distance between itself and request sender node. Nodes that receive the responses from its neighbour nodes calculate the value of willingness to join the cluster μ as follows;

Definition 5: *Let di is the distance between sender and receiver, dc is the distance between cluster head and sender and h is the number of hops between cluster head and sender*

$$\mu = 1 - \left(\frac{1}{(di + dc) \times (h + 1)}\right)$$

When a cluster head accede the number of connection (connection limit λ it broadcast a new cluster creation packet to its cluster member. Each node that receives cluster creation packet, respond back to cluster-head with its geographical location and number of neighbouring nodes. The cluster-head after receiving the responses decide the new cluster-head as follows;

Definition 6: *Let N is set nodes that responded back to the cluster head, d is set of their distance with cluster-head locations, and η as set of number of neighbours we define the function of selection π for each node as follows;*

$$\pi = \frac{\eta}{\lambda} \times d$$

Definition 7: *Let Ω is the function for selecting a cluster-head from a set of cluster-heads we define Ω as follow;*

$$\Omega(\mu_1, \mu_2 \ldots \ldots \mu_N) = \max\ (\pi_1, \pi_2 \ldots \ldots \pi_N)$$

The last step for cluster construction is merging two clusters. When a cluster member start leaving the cluster-head and cluster member go below a specific threshold, ρ, Cluster-head becomes interested in join the large cluster from its neighbouring clusters. The cluster-head broadcast the packet in the network to get the information about other clusters in the network. When this broadcasted packet is received by the cluster-member of other another cluster they forward the request to their respective cluster-head. In both case, either cluster-head receive the request packet by itself or receive through its membership function it compute the feasibility of merging clusters by check the condition (current members + new members < threshold value). If the condition satisfies the cluster allow the small cluster to merge its members.

5 Experiments and Results

For our experiments we use NS2 as network simulation. We wrote our Data aggregation algorithm as a NS2 application/Agent that was associated with all nodes We first generated cluster based environment using NS2 random walk wireless model. When the simulation was run node generated data and start building clusters. We perform experiments in Ubuntu Linux environment with NS2.35 simulator. The computer that we used for our experiments was Core i5 with 4 GB RAM. Our observation includes the size of data that was dissimilated on the network, number of clusters and their influence on data aggregation technique. However we didn't examine the bandwidth specifically but we examine the data size instead of that.

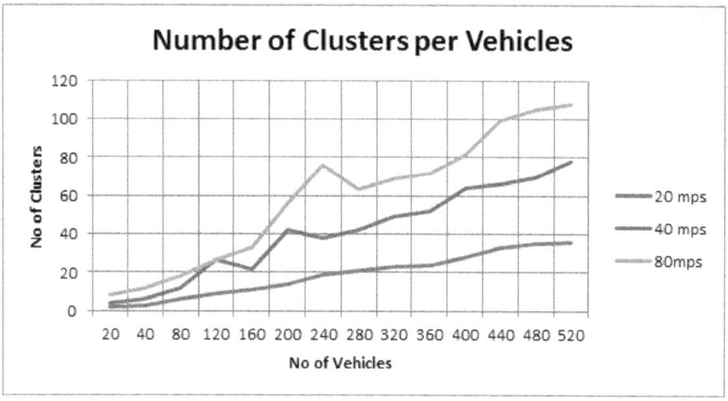

Fig. 2. Number of Clusters and Cluster vehicle relationship

Figure 2 shows the number of created clusters for different number of vehicles and mobility effect on the cluster creation environment. This figure shows that number of clusters increased while we increased the speed of vehicles. This shows that if the traffic will be slow the clustering will be more efficient then the clustering with high speed. This is because of the mobility in Vehicular Network as fast as the vehicle will move the probability of leaving a cluster and finding a new cluster will increase. Thus number of clusters at high speed highway will increase. Our experiments also shows that in rush the number of clusters becomes ideal where one cluster handle about 25-30 vehicles on 4 lane bidirectional road segment

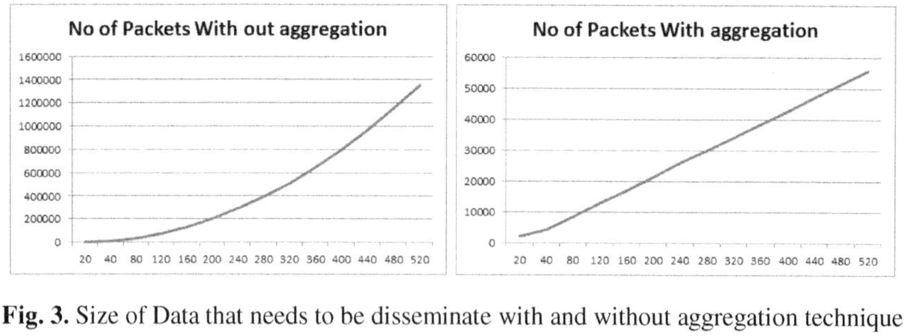

Fig. 3. Size of Data that needs to be disseminate with and without aggregation technique

Figure 3 shows the success of our aggregation technique by depicting the number of packets using cluster based aggregation technique and Zero Aggregation technique. The graph in Figure 3a shows the number of data packets without aggregation. To ensure the results we run this simulation 5 times and the presented results are average results of 5 performed aggregation. The graphs clearly show the decreasing numbers of packets in figure 3b where we applied our aggregation technique.

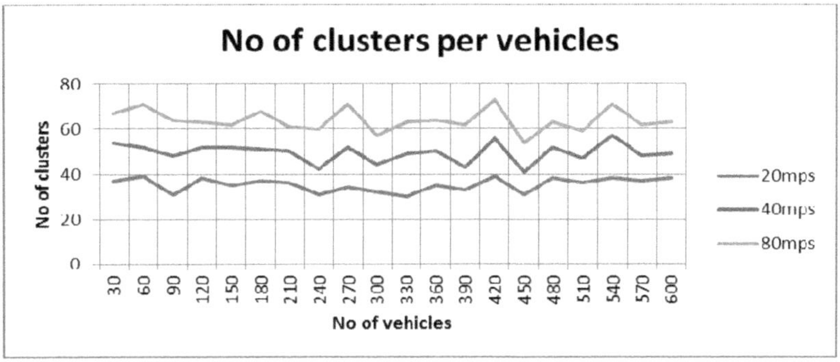

Fig. 4. Number of Clusters for a specific road segment for 10 minutes

Figure 4 shows the varying behaviour of clusters with time of 500 vehicles. The simulation was run for 600s i.e. 5 mints to observe the behaviour of the clusters during the moving traffic to examine the mobility prospective of the clustering algorithm. This was important because the whole concept of aggregation and dissemination is based on clustering algorithm. Thus the cluster creation, formation and maintenance will be, this put direct effect on the aggregation technique. This is because the ration with which cluster increase in an VANET environment, with the same ration number of packets increases.

6 Conclusion

This paper has proposed a new data aggregation technique for vehicular ad hoc networks based on combination of functional equation based information fusion and autonomous clustering. In propose approach each vehicle becomes part of some cluster while it enters on a road segment. The vehicle sends its recoded sensor data to the cluster-head for aggregation. The cluster head receives data from its cluster members and perform aggregation operation periodically with a define time. Once cluster head complete performing aggregation it disseminates the aggregated data to its members. For formation and maintenance of clustering in VANET environment we have used autonomous clustering technique. The experimental results show that aggregation technique can successfully reduce the size of data for dissemination over the network that was the core focus of this work.

Acknowledgements. This research was supported by Basic Science Research Program through the National Research Foundation of Korea (NRF) funded by the Ministry of Education, Science and Technology (2011-0012329).

References

1. Yu, B., Gong, J., Xu, C.-Z.: Catch-up: A Data Aggregation Scheme for VANETs. In: VANET 2008: Proceedings of the Fifth ACM International Workshop on VehiculAr Inter-NETworking, pp. 49–57. ACM, New York (2008)
2. Wischhof, L., Ebner, A., Rohling, H., Lott, M., Halfmann, R.: SOTIS - a self-organizing traffic information system. In: Proc. VTC 2003-Spring Vehicular Technology Conference The 57th IEEE Semiannual, April 22–25, vol. 4, pp. 2442–2446 (2003)
3. Caliskan, M., Graupner, D., Mauve, M.: Decentralized Discovery of Free Parking Places. In: Proceedings of VANET 2006, pp. 30–39. ACM, New York (2006)
4. Bae, K., Yoon, H.: Autonomous Clustering Scheme for Wireless Sensor Networks Using Coverage Estimation-Based Self-Pruning. Special Section on Ubiquitous Networks, IEEE Transaction on Communication E88–B(3), 973–980 (2005)
5. Dietzel, S., Bako, B., Schoch, E., Kargl, F.: A Fuzzy Logic based Approach for Structure-free Aggregation in Vehicular Ad-Hoc Networks. In: Proc. VANET 2009, Beijing, China, September 25. ACM (2009)
6. Lochert, C., Scheuermann, B., Mauve, M.: Probabilistic aggregation for data dissemination in VANETs. In: VANET 2007: Proceedings of the Fourth ACM International Workshop on Vehicular Ad Hoc Networks, pp. 1–8 (September 2007)
7. Ohta, T., Kawaguchi, T., Kakuda, K.: A New Multicast Routing Protocol Based on Autonomous Clustering in Ad Hoc Networks. In: Proc. Autonomous Decentralized Systems, ISADS 2005, pp. 297–305. IEEE (2005)
8. Quan, S.G., Kim, Y.Y.: Fast Data Aggregation Algorithm for Minimum Delay in Clustered Ubiquitous Sensor Networks. In: Proc. International Conference on Convergence and Hybrid Information Technology, ICHIT 2008, pp. 327–333. IEEE (August 2008)
9. Rathod, M., Mahgoub, I., Slavik, M.: A hybrid data dissemination scheme for VANETs. In: Proc. Wireless Days (WD). IFIP, pp. 1–7. IEEE (2011)
10. Dietzel, S., Kargl, F., Heijenk, G., Schaub, F.: Modeling in-network aggregation in VANETs. In: Proc. IEEE Communications Magazine. IEEE (2011)

Simulations of VANET Scenarios with OPNET and SUMO

Florent Kaisser, Christophe Gransart, and Marion Berbineau

Univ Lille Nord de France, F-59000, Lille,
IFSTTAR, LEOST, F-59650, Villeneuve d'Ascq
{florent.kaisser,christophe.gransart,marion.berbineau}@ifsttar.fr

Abstract. Vehicular Ad hoc Networks (VANET) are a special kind of Mobile Ad-Hoc Networks (MANET) adapted to the communications between vehicles. Several specific protocols to VANETs have been developed to improve performances and satisfy vehicular application needs. To evaluate a protocol for VANET, some realistic mobility models are needed. Unfortunately, such models are not provided by Opnet Modeler. In this work, we propose a framework that enhances Opnet Modeler simulation scenario using realistic vehicular mobility models. This framework makes use of the open source software called "Simulation of Urban MObility" (SUMO) and the "input trajectory files" feature of Opnet Modeler.

Keywords: VANET, Opnet, SUMO, simulation, framework, dissemination.

1 Introduction

Starting with the idea of making driving safer by inter-vehicle communication, the concept of vehicular networks or vehicular ad hoc networks (VANETs) has been extended to a large collection of various applications that can profit from wireless communication between vehicles. Vehicles are not only communicating between each other, but are also getting information from and sending data to infrastructure units. This particular context has several specificities like the deployment of safety applications that have critical requirements regarding communications. Classical ad hoc routing protocol (AODV, OLSR, GPSR) are usable in VANET, but they are not adapted to the specificities of these networks.

Several specific protocols to VANETs have been developed to improve performances and satisfy vehicular application needs. To evaluate scalability and efficiency of a protocol for VANET, some simulations with realistic mobility model are needed. Numerous works use various tools to evaluates their proposals. In this paper we propose to use Opnet Modeler to simulate VANET by associating it with a vehicular mobility simulator.

In a first section, we propose an overview of VANET simulation by describing current simulation tools. Afterwards, in a second section, we introduce our

A. Vinel et al. (Eds.): Nets4Cars/Nets4Trains 2012, LNCS 7266, pp. 103–112, 2012.

framework to simulate VANET with Opnet Modeler. Finally, we use our work to evaluate a simple dissemination protocol for VANET by simulation of adapted scenarios.

2 VANET Simulation Overview

In this section, we give a brief overview of vehicular ad hoc network and tools to simulate them. Like other kind of networks, simulating VANET is necessary to validate protocols and applications. In addition, the total number of vehicles in a network can be large, so evaluating the scalability is central.

Many tools allow to simulate VANET. In many case, this tool is a vehicular mobility model associate with a network simulator. In some case, vehicular mobility model is a simple random way point mobility model, but it restricts the movement of vehicles to real roads. To improve realism of the simulation, a car-following model can be added, allowing to compute the vehicle acceleration (then speed and position) according to neighbor vehicles in the same lane or adjacent lanes. Next, we describe some network simulators associated to a vehicular mobility model. A more exhaustive list of tools to simulate VANET is depicted in a dedicated book [1] and a paper [2].

SWANS++ [3] is the network simulator SWANS (from JiST/SWANS [4] project) extended with a vehicular mobility model, STRAW [5] (Street Random Way point). STRAW uses a simple random way point mobility model but it takes in account a real street map loaded from TIGER [6] (Topologically Integrated Geographic Encoding and Referencing, an US street database) data files.

GrooveNet [7] (originally known as GrooveSim) is a network and mobility simulator for real and simulated vehicles. GrooveNet, originally as an extension of RoadNav [8] open-source simulator, can load real street map from TIGER database and includes a car-following model.

TraNS [9] (Traffic and Network Simulation Environment) was the first work to combine the network simulator ns-2 [10] with a vehicular mobility simulator SUMO. We introduce SUMO below. The link between the two simulators is done with a parser that reads the output of SUMO and converts it to a suitable format for ns-2. Other works like MOVE [11] GUI and *traceexporter* tool (module of SUMO) allow to link ns-2 with SUMO.

In Veins [12] (Vehicles in Network Simulation), SUMO is paired with an other network simulator OMNET++ [13]. It creates a bidirectional communication between SUMO and OMNET++. Thus, the network simulator can react to events from the mobility simulator.

iTETRIS simulation platform [14] use ns-3 [15] as network simulator and SUMO. The two open source simulation platforms are connected with a central module called iTETRIS Control System (iCS). This approach permits to have the best performance in terms of scalability and modularity.

Opnet Modeler enables both random mobility and trajectory mobility. The last-mentioned is setting with one *trajectory file* per node. This method allows to elaborate complex node movements. Unfortunately, interaction between nodes are not taken in account, so realistic mobility is limited. To the best of our knowledge, no tool for Opnet Modeler exists allowing to simulate VANET with a realistic mobility model.

In this paper we propose a framework for Opnet Modeler to simulate complex mobility, in our case moving vehicles on a road, using an extern microscopic traffic simulation called SUMO.

3 A Framework to Simulate VANET

A microscopic traffic simulator computes the position of all vehicles on road at some time t from vehicles and road characteristics. These positions are used to show vehicles on a graphical user interface (GUI) or written in a file (*dump file*). These simulators use a realistic model to describe the vehicles moving. This model computes the ideal acceleration, velocity and position for each vehicle any time, therefore avoid the physical collisions between vehicle.

Input data can be initials positions, vehicle characteristics (acceleration, maximum speed, etc), road network,way to go, position of destination, speed limits, etc. SUMO [16] is a free implementation of such a simulator and supports car-following model (issue of [17,18,19,20]). SUMO implementation consists of several modules (or commands), with the SUMO module in the center (Figure 1). Input data of SUMO is one or several *road network files* and one route file. The road network file can be generated from various input formats with netconvert module. These input formats can be *osm files*, from Open Street Map [21] database, or *shape files*, from TIGER [6] database. Many other formats are available.

A *route file* describes both the characteristics of vehicles on the road and the path taken by the cars (edge sets of network road). Like road network files, route files can be generated with a other module called *duarouter* from two kinds of input: *trip* or *flow* definition. Each trip consists at least of the starting, the ending edge and the departure time. Then *duarouter* computes the shortest path between theses edges. Each trip is associated with one vehicle. Flow is mostly the same approach as using trip definitions, but one may join vehicles having the same departure and arrival edge using this method. Moreover, initial position of all vehicles are computed randomly (from a simple discrete uniform distribution).

3.1 Generate Simple Scenario

From network and route files, SUMO module simulates all the positions at several times (e.g. every seconds). These positions are written in a dump file. This file is used to generate file(s) in a format understood by network simulator softwares. In our case, we need a converter to generate files for Opnet Modeler. This is achieved by developing our extension *traceexporter* [1]. This extension implement

[1] Extension *traceexporter* already exists, but it only for ns-2. We have preferred re-develop a similar extension for both ns-2 and Opnet Modeler simulators. Moreover our implementation is more modular and therefore more easier to extend.

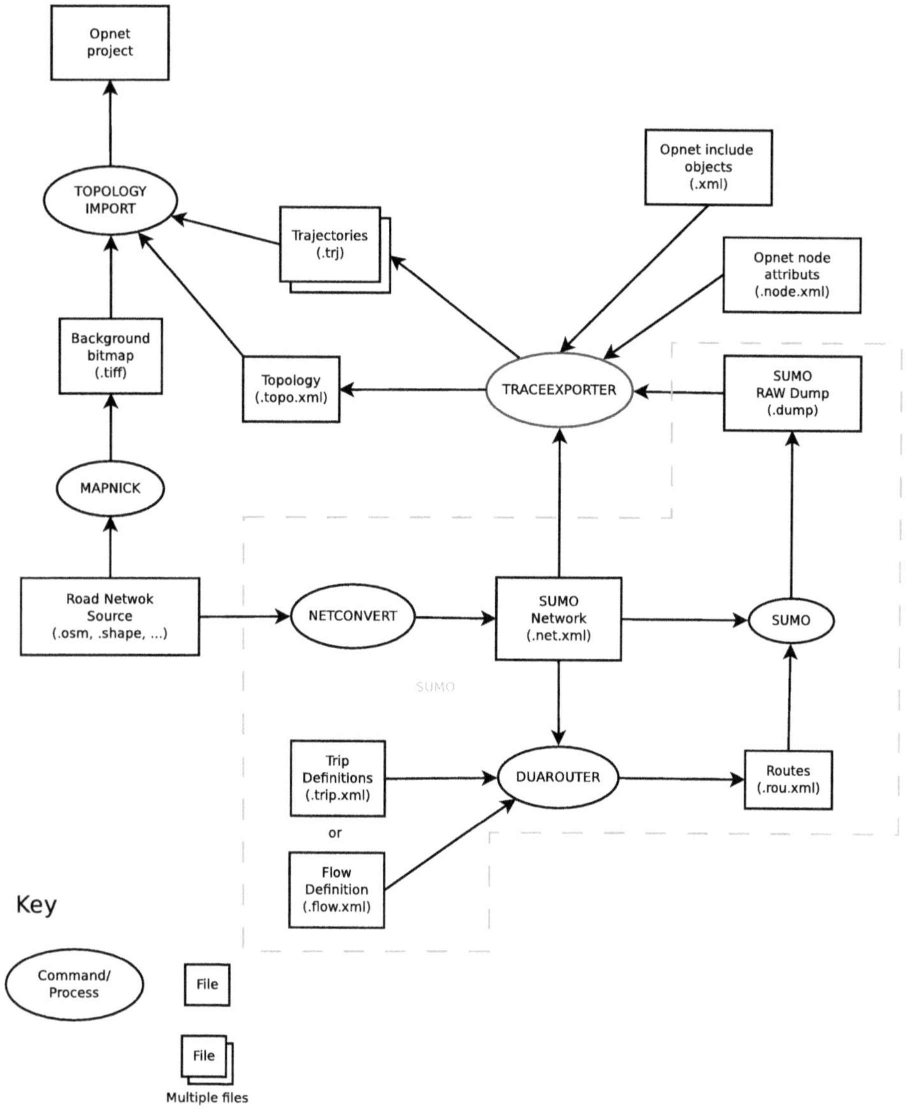

Fig. 1. Process to generate vehicular mobility with SUMO

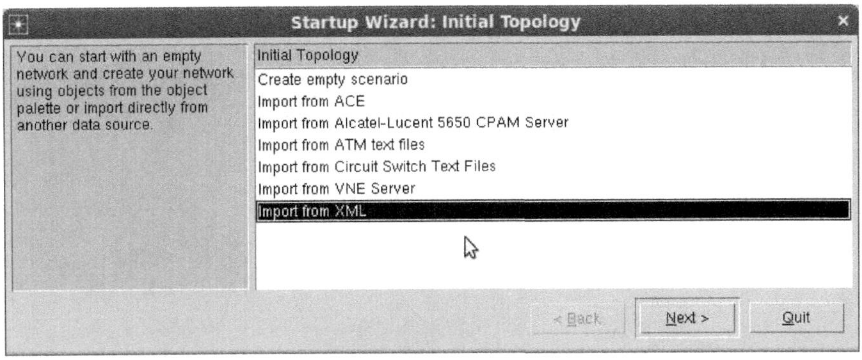

Fig. 2. Topology import with new scenario assistant

the data flow diagram shown in Figure 1. In addition of dump file, two other input XML files are added: node attributes and include objects. The scheme of these XML files are the same as the files generated by Modeler exporter feature. The node attribute file contains the attributes of each vehicle node (model name, parameters, icon, etc). The include object file contains the description of all objects to be added to the scenario, e.g. a backbone or statics equipments.

traceexporter generates one *trajectories file* for each vehicle and one *topology file*. The trajectories file is the same file generated by the GUI of Modeler when user defines a trajectory for a node. Topology file contains the scenario description with nodes description (name and initial vehicles position from SUMO dump file, and attributes from *node attribute file*) and other objects description (from include *object file*). In addition of this file, a background bitmap can be generated from network road file with adequate tool (e.g. *mapnick* for osm file).

Once trajectories files are copied to a Opnet model directory, the user calls the Topology importer feature from Opnet Modeler (Topology menu, Topology import, from XML files...), or creates a new scenario with XML import option (Figure 2). User should obtain a results like shown in Figure 3. Discrete Event Simulation can be run from this point.

3.2 Generate Multiple Scenarios

We added a feature to *traceexporter* allowing to generate multiple scenarios at the same time. A *scenario file* defines all scenarios to generate. Each scenario contain links to one input file for the road network (road network file), one for the route definitions (route file, trip file or flow file), one for the include objects and one for the node attributes (Figure 4). The different scenario can use same or different input files. For examples, to generate scenarios with different number of vehicles (50, 100, 150), three scenarios use the same network and node attributes files, but three different flow files containing three number of vehicles to generate. Moreover, *traceexporter* runs automatically the commands like SUMO and *duarouter* (for this reason, SUMO does not appear on Figure 4).

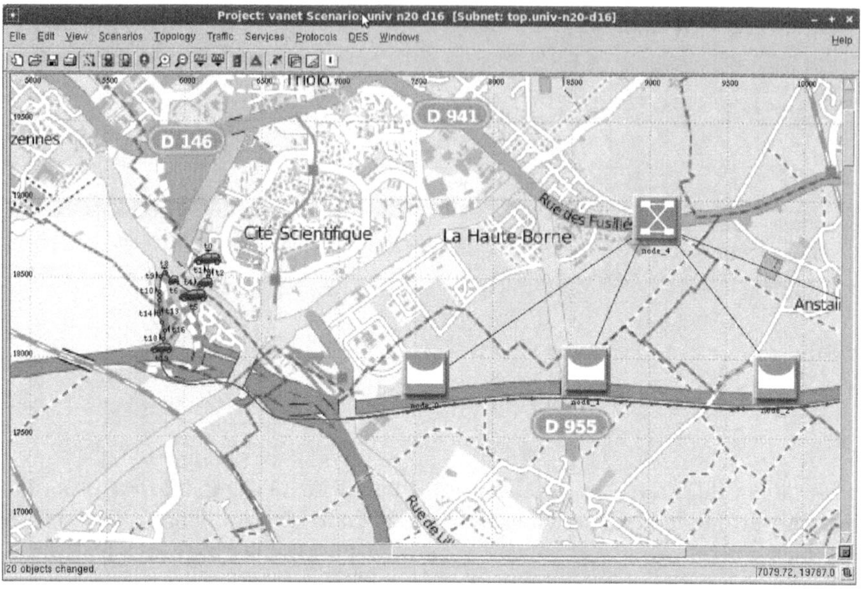

Fig. 3. VANET in Opnet Modeler with our framework

After execution of *traceexporter*, one topology file is generated for each scenario in different directories with their trajectories files associated.

4 Dissemination of Informations in VANET

Several applications for VANET need to disseminate information on the network. For security, an alert is spread to neighbor vehicles or the current informations of vehicles (like the position, speed or other information from vehicle probes) can be sent to all neighbors to increase driver visibility. We take this application as an instance of a VANET simulation. Therefore, we add a process model for Opnet Modeler to simulate a dissemination of information in VANET. This process is above *wireless_lan_mac*. A new packet format is defined, containing: vehicle ID, type of information, position, number of hop, lifetime of the information, sequence number. This packet is the payload of a MAC packet sent by MAC layer. The process sends a packet every 500 ms. We define a simple forwarding scheme (flooding): when the process receives a packet not received in the past, the packet is re-sent to the neighbors. Thus, all nodes in the network receive informations contained in the packet.

With a high density of vehicles, the number of packet sent in the network can reach the network capacity, packets are no longer forwarded, then some vehicles do not receive information. To avoid this, we define a forwarding scheme to limit the number of packet sent. Several existing works [22,23,24,25] propose an achievement of dissemination in VANET. In this paper, we simulate (in addition to

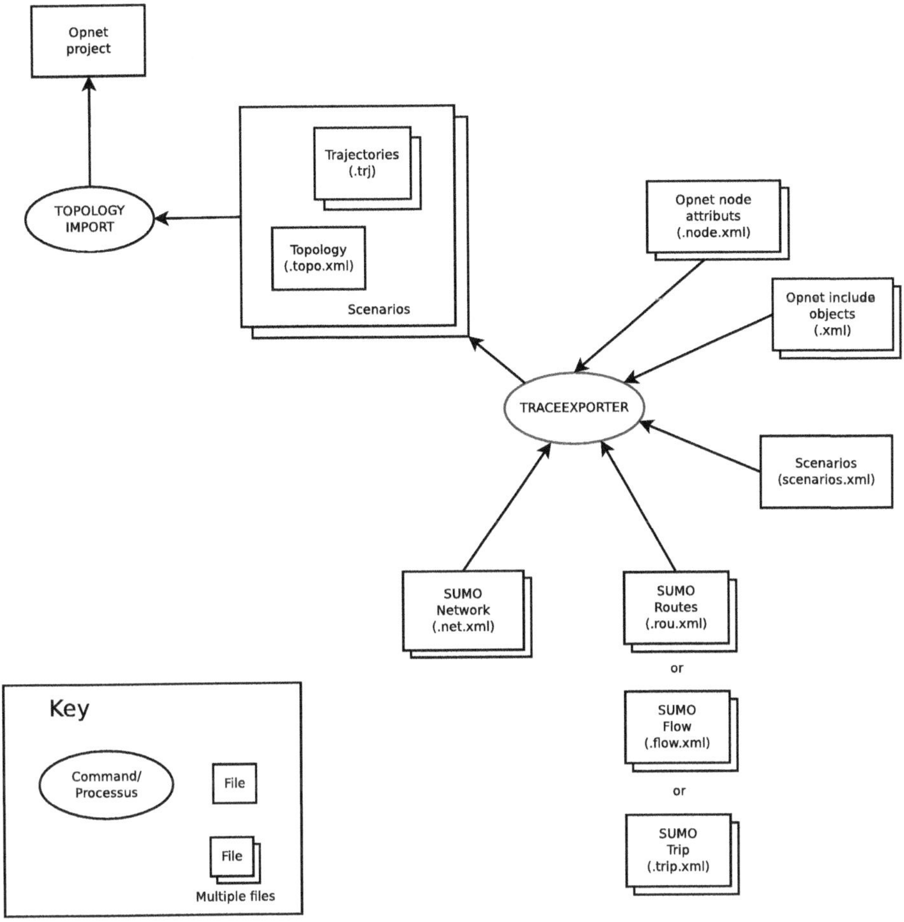

Fig. 4. Multiple scenarios generation with *traceexporter*

flooding) a simple distance-based protocol (SDP) [25]. On message reception, a vehicle starts a timer T inversely proportional to the distance from previous sender:

$$T = t_{max} * (1 - \frac{d}{R})$$

with t_{max} the maximum waiting time, d the distance from previous sender, R the radio range. In the simulation, $R = 550\ m$ and $t_{max} = 50\ ms$. When the timer reaching zero the node forward the packet. Other waiting nodes, receive this packet and cancel their timer. Thus, we reduce the number of packet sent in the network. We simulate flooding and SDP protocol with several number of vehicles. In Figure 5 we see a good improvement with SDP protocol: the maximum number of discovered neighbor with flooding is about 120, whereas the maximum is close to 200 with SDP.

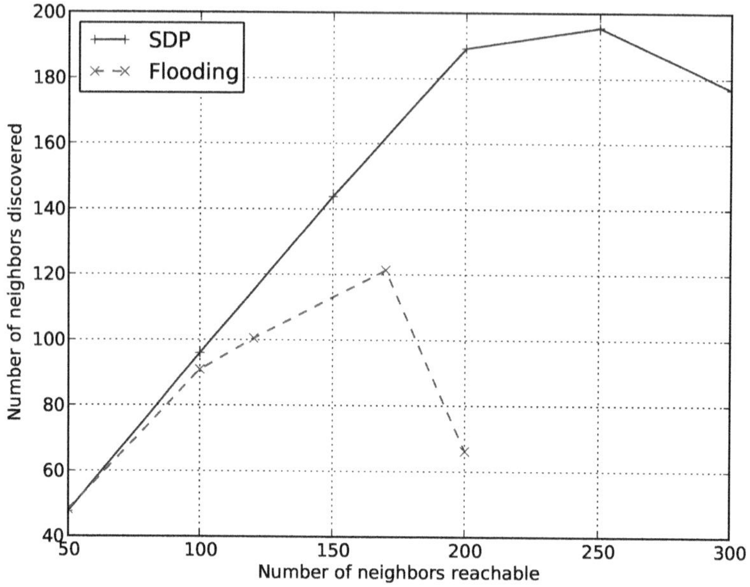

Fig. 5. Neighboor discovery with simple dissemination protocols

5 Conclusion

In this paper we have proposed a framework to simulate VANET from Opnet Modeler and SUMO. Our proposal allows to use a realistic mobility model to compute movement of vehicles on a real road with Opnet Modeler to effectively simulate communication networks and with many existing models. The aim of this framework is to be modular and easy to use by exploiting judicious feature of Modeler like topology import.

But simulation of VANET with several number of vehicles or random seed require many Opnet Modeler scenarios and an equivalent number of topology import, which limits a good protocols evaluation. In future works, we wish to integrate in Opnet Modeler a direct and bidirectional communication with SUMO allowing us to define more complex simulation scenarios.

Source code of this work can be downloaded at `https://bitbucket.org/florent_k/traceexporter`

References

1. Olariu, S., Weigle, M.C.: Vehicular Networks: From Theory to Practice. Chapman & Hall and CRC (2009)
2. Martinez, F.J., Toh, C.K., Cano, J.C., Calafate, C.T., Manzoni, P.: A survey and comparative study of simulators for vehicular ad hoc networks (vanets). Wireless Communications and Mobile Computing 11(7), 813–828 (2011)

3. SWANS++: Extensions to the scalable wireless ad-hoc network simulator, http://www.aqualab.cs.northwestern.edu/projects/swans++/ (Online; accessed January 2, 2012)
4. Jist/Swans: Java in simulation time/scalable wireless ad hoc network simulator, http://jist.ece.cornell.edu (Online; accessed January 2, 2012)
5. Choffnes, D.R., Bustamante, F.E.: An integrated mobility and traffic model for vehicular wireless networks. In: Proceedings of the 2nd ACM International Workshop on Vehicular Ad Hoc Networks, pp. 69–78. ACM (2005)
6. TIGER: Topologically integrated geographic encoding and referencing, http://www.census.gov/geo/www/tiger/ (Online; accessed January 2, 2012)
7. Mangharam, R., Weller, D.S., Stancil, D.D., Rajkumar, R., Parikh, J.S.: Groovesim: A topography-accurate simulator for geographic routing in vehicular networks. In: Proceedings of the 2nd ACM International Workshop on Vehicular Ad Hoc Networks, pp. 59–68 (2005)
8. Roadnav, http://roadnav.sourceforge.net (Online; accessed January 2, 2012)
9. Piorkowski, M., Raya, M., Lugo, A.L., Papadimitratos, P., Grossglauser, M., Hubaux, J.P.: Trans: realistic joint traffic and network simulator for vanets. ACM SIGMOBILE Mobile Computing and Communications Review 12(1), 31–33 (2008)
10. NS2: the network simulator, http://www.isi.edu/nsnam/ns (Online; accessed January 2, 2012)
11. Karnadi, F.K., Mo, Z.H., Lan, K.: Rapid generation of realistic mobility models for VANET. In: IEEE Wireless Communications and Networking Conference, WCNC 2007, pp. 2506–2511 (2007)
12. Sommer, C., Yao, Z., German, R., Dressler, F.: Simulating the influence of ivc on road traffic using bidirectionally coupled simulators. In: IEEE INFOCOM Workshops 2008, pp. 1–6. IEEE (2008)
13. OMNeT++ community site, http://www.omnetpp.org (Online; accessed January 2, 2012)
14. itetris test-beds, http://www.ict-itetris.eu/platform.htm (Online; accessed January 2, 2012]
15. The ns-3 network simulator, http://www.nsnam.org (Online; accessed January 2, 2012)
16. SUMO - Simulation of Urban MObility (2012), http://sumo.sourceforge.net (Online; accessed January 2, 2012)
17. Krauß, S.: Microscopic modeling of traffic flow: Investigation of collision free vehicle dynamics (1998)
18. Wagner, P., Lubashevsky, I.: Empirical basis for car-following theory development. Arxiv preprint cond-mat/0311192 (2003)
19. Kerner, B.S.: Introduction to modern traffic flow theory and control: the long road to three-phase traffic theory. Springer (2009)
20. Treiber, M., Hennecke, A., Helbing, D.: Congested traffic states in empirical observations and microscopic simulations. Physical Review E 62(2), 1805 (2000)
21. Haklay, M., Weber, P.: Openstreetmap: User-generated street maps. IEEE Pervasive Computing 7(4), 12–18 (2008)
22. Kim, T.-H., Hong, W.-K., Kim, H.-C.: An Effective Multi-hop Broadcast in Vehicular Ad-Hoc Network. In: Lukowicz, P., Thiele, L., Tröster, G. (eds.) ARCS 2007. LNCS, vol. 4415, pp. 112–125. Springer, Heidelberg (2007)

23. Briesemeister, L., Schafers, L., Hommel, G.: Disseminating messages among highly mobile hosts based on inter-vehicle communication. In: 2000 IEEE Intelligent Vehicles Symposium, Dearborn, MI, USA, pp. 522–527 (2000)
24. Sun, M.-T., Feng, W.-C., Lai, T.-H., Yamada, K., Okada, H., Fujimura, K.: Gps-based message broadcasting for inter-vehicle communication. In: Proceedings 2000 International Conference on Parallel Processing, Toronto, Canada, pp. 279–286. IEEE Computer Society (2000)
25. Hrizi, F., Filali, F.: Achieving broadcasting efficiency in v2x networks with a distance-based protocol. In: 2009 IEEE Communications and Networking (ComNet), Hammamet, Tunisia, pp. 1–8 (2009)

Human Participatory Sensing in Fixed Route Bus Information System

Bhushan G. Jagyasi, Vikrant Kumar, and Arun Pande

TCS Innovation Labs Mumbai, TATA Consultancy Services, India
{bhushan.jagyasi,vikrant4.k,arun.pande}@tcs.com

Abstract. In urban areas, a large population relies on buses with fixed route for commuting within a town. However, in most of the developing and underdeveloped countries, real-time bus arrival time information is still unavailable to the commuters. This is mainly due to the difficulty in installation and maintenance of existing infrastructure based solutions like Global Positioning System or Radio Frequency Identification system. In the present paper, we propose a human participatory instantaneous bus location information system, which relies on the ubiquity of mobile phones and on human participation in an opportunistic manner. We further propose an algorithm to estimate bus arrival time at different stops on the route. The localization and arrival time information is then provided to the commuters, either in response to their query or in the form of alerts. We believe that the proposed system can be made self sustainable by providing appropriate incentives to the commuters in lieu of localization of the buses. The provision of real time bus information can become an important incentive.

1 Introduction

Road traffic in most urban areas is very dynamic in nature due to increasing number of vehicles, varying weather conditions, travel time of the day and events on the road. Eventhough the scheduled starting time from the originating stop is fixed, there is substantial uncertainty in the arrival time at subsequent stops on the route. This uncertainty affects the convenience and the traveling experience of the commuters relying on the public transport system. There is hence a need for a system that provide: (a) localization information of bus trips on the fixed route and (b) the expected time the bus may take to arrive at a selected stop where a passenger is supposed to board.

Most of the existing approaches ([1,2,3,4]) are based on the installation of a Global Positioning system (GPS) and a wireless communication device on buses to track the instantaneous vehicle location. Apart from the huge deployment cost of on-board GPS-based systems, GPS enabled hand-held devices have high power consumption. Further, GPS enabled devices results in inaccurate readings due to lack of visibility to GPS satellites and they may need complex algorithms to estimate the arrival time. Because of these drawbacks, most of the GPS based approaches have not yet been deployed in the urban ares of developing and underdeveloped countries.

A. Vinel et al. (Eds.): Nets4Cars/Nets4Trains 2012, LNCS 7266, pp. 113–123, 2012.

Other existing approaches [5], [6] for tracking fixed route buses are based on deployment of Radio Frequency Identification (RFID) reader along with a wireless communicator on the buses and the active RFID tags on the bus stops. The system helps to determines the time of arrival of a bus at a particular stop and communicate the same to the central server. This system is also not widely deployed because of the huge infrastructure cost and the errors due to communication losses. It further requires periodic maintenance of the system and provision of continuous power to all RFID active tags on the bus stops. Recently, Quick Response (QR) codes are being used in Paris and Tokyo in central bus stations to provide arrival time information to passengers at bus stops [7]. Each bus stop is assigned a unique 2-D QR code, which can be scanned through the smart phones to get the information of the subsequent buses on that bus stop. This system [8], is hence a cost-effective alternate solution to replace the information displays on the bus-stops.

In recent literature, Human participatory sensing [9], [10], [11], [12], [13], [14], [15] has been used in different applications of urban sensing like urban planning, market price information, health and environment. Here, we explore ubiquitous mobile phones for transport application in order to provide arrival time information of the fixed route buses in any metropolitan city in a developing country.

In this paper, we envisage the update of bus localization information with the help of participation from the passengers traveling on the same bus using a mobile phone application. This indeed is a *mobile phone based human participatory sensing*, wherein human beings use their own senses to observe the physical phenomena and communicate the same using mobile phone. This distributed and participatory localization approach results in a real time dynamic database as the buses move in the town. This dynamic database containing Bus location information at each stop is further utilized to provide the bus traveling coordinates and its arrival time information at a particular bus stop in response to the query from a mobile phone or a web console. The key contribution of this paper is the novel human participatory sensing framework for real time bus information systems. We also propose a bus arrival time estimation algorithm based on the data obtained from the proposed human participatory sensing framework.

Section 2 presents the schematic of the proposed human participatory bus arrival time information system. The proposed bus arrival time estimation algorithm based on participatory sensed data is provided in Section 3. The experimental results are presented in Section 4. Finally, Section 5 concludes the paper and presents our future work directions.

2 Framework for Human Participatory Sensing Based Dynamic Bus Arrival Time Information System

Unlike existing infrastructure based approaches, we here propose to utilize the existing resources of human beings traveling on buses. We consider humans carrying mobile phones as moving sensors nodes. In this paper, we exploit the power of human participatory sensing to obtain the status of buses in urban areas.

The schematic of the proposed human participation based bus arrival time information system is shown in Fig.1. Without loss of generality, we have developed a mobile phone application on the Android platform which facilitates a person traveling on the Bus to update the current location in an user-friendly manner. This can be done by just entering the Bus number at the start of the travel and then selecting the current bus stop from the list of bus stops as bus arrives at a particular stop. This dynamic information about the current location of the bus gets updated using the GPRS service of the mobile service provider, usually available in developing countries. Further note that although the application has been developed for the Android mobile platform, it can be developed on any low-end mobile phone with the same features.

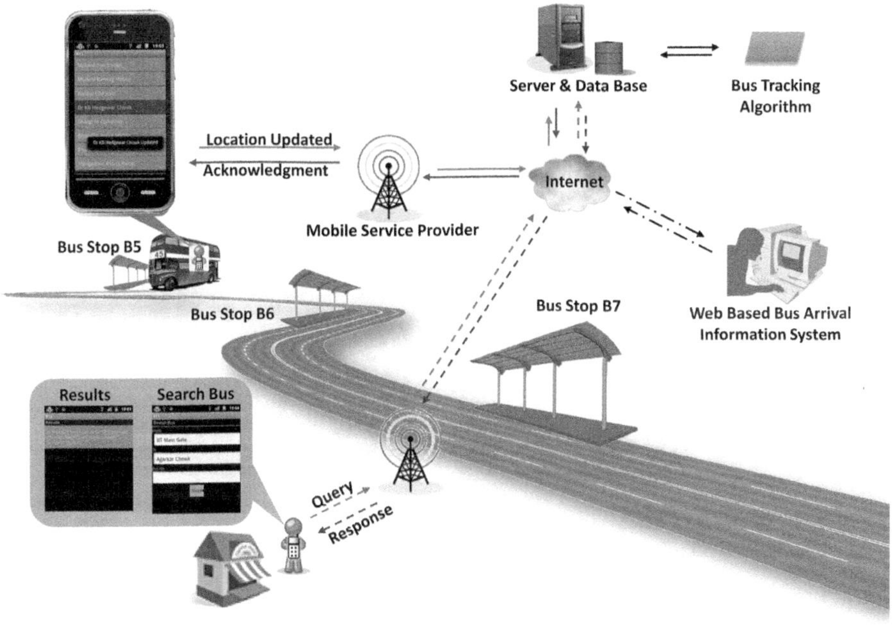

Fig. 1. Human Participatory Sensing for Bus Information System

On the update of any bus stop location, the proposed bus arrival time estimation algorithm, presented in Section 3, computes the arrival time at all the subsequent bus stops and stores the same in the database. The mobile phone application also permits the user waiting for a bus to obtain the status of the running bus along with its estimated arrival time at any subsequent stop on the route. As in Fig. 1, a person close to the stop B7 taps a query for finding the Bus from the desired location to the destination. On submission of the query, based on the past statistics, current running conditions, and current bus location, the list of buses arriving sequentially along with their estimated arrival time and last seen information is provided by the mobile phone application (Fig. 2). Taking into consideration human error, even if the bus location information for some stops gets missed out, the bus arrival time information system works on the available limited data set to obtain the estimate of the bus arrival time information and still one gets information on when and where was it last seen.

As seen in Fig. 3, the application also permits the users to get the real time status of the buses along with their arrival time information at any bus stop using the Google map application program interface based web application (Fig. 3). The user, sitting in an office or at home, can enter the bus number and desired direction of travel. This instantly triggers the map and shows the current location of all buses on the selected route on the map. On selecting the desired boarding point, the estimated arrival time and the localization information of the nearest bus gets displayed. This significantly reduces the anxiety and the waiting time on the bus stops.

Fig. 2. Query and update of Bus Information on Mobile Phone application

In a city where bus travel time has a great variability due to a number of factors, it becomes important for passengers traveling on a bus to have information about when the bus will reach the destination. Hence, along with getting information of a bus arrival time at an originating stop, the users on the same mobile phone application can also query on when they shall reach their destination stop while traveling on a bus.

Throughout this paper, we assume that there is atleast one person on the bus or at the bus stop who updates the bus localization information based on their observations. With the availability of user friendly means to update the bus location information, this task can also be taken care by the bus crew member in case of no passenger volunteering to do that. This method overcomes the drawback of existing approaches and make utilization of the available resources in a cost

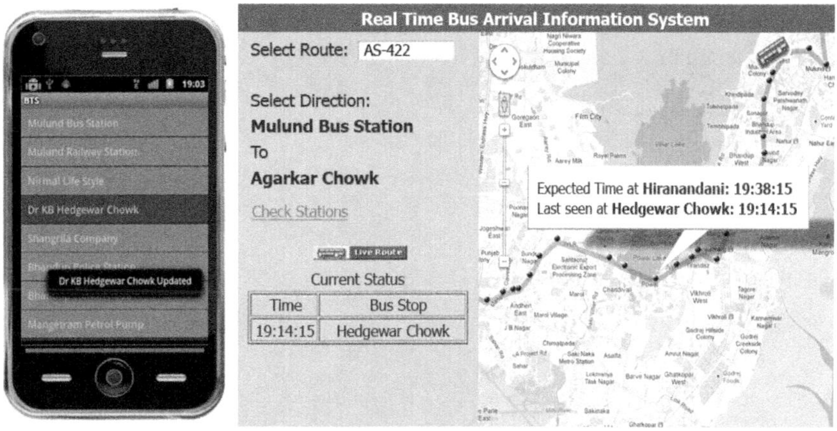

Fig. 3. Real time bus update on web-console

effective manner. It neither requires deployment of any additional infrastructure nor incurs maintenance cost.

Since the proposed architecture is based on the crowd sourcing, an appropriate incentive mechanism has to be devised. This topic is beyond the scope of the paper. However, we suggest that, in an urban scenario of a developing country, an unpaid service of real time arrival information on itself becomes a good incentive.

3 Proposed Arrival Time Estimation Algorithm

In Section 2, the human participatory bus information system has been presented which enables bus localization in order to obtain the instantaneous time of buses at any bus stops on the route. In this section, we present a bus arrival time estimation algorithm to facilitate the commuters with the arrival time information at any desired stop. The proposed algorithm not only relies on the history of the bus arrival information, but also on its current running status.

The time taken by a bus to travel on a particular route in Mumbai (refer Section 4 for experimental setup) for different starting time segments has been shown in Fig. 4. As seen in the figure, there is high variance seen in the bus travel time which indicates different traffic conditions on the same route for the different time slots. However, for a same time slot, on a weekday in the subsequent part of this paper, and for a given season, similar pattern of travel time has been observed. This motivates to obtain a separate *arrival time estimate* for different segments of the time slots, independently.

We consider a bus route b with a fix number of stops M having sequence number $m \in \{1, 2 \cdots, M\}$. The travel time estimation for any two consecutive stops is obtained for each time segment of the day. This requires segmentation of twenty four hours of a day into N number of discrete time slots having time slot index $n \in \{0, 1, \cdots, N-1\}$. If a time slot duration is one hour then $N = 24$ and $n \in \{0, 1, \cdots, 23\}$ corresponds to the time slots $\{(00 : 00) - (00 : 59),$

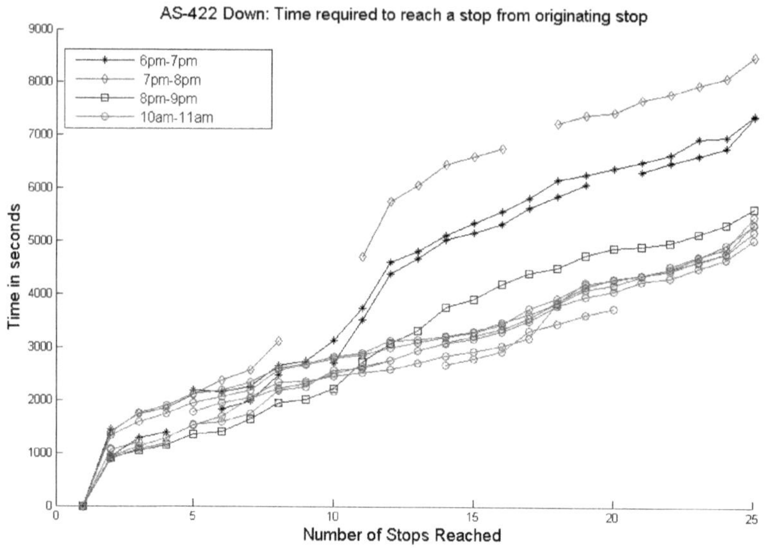

Fig. 4. Time taken by Bus to reach different stops on the same route for starting time in different slots

$(1:00)-(1:59),\cdots,(23:00)-(23:59)\}$, respectively. The stop sequences i and j are the adjacent stop sequences with $j=i+1$, and T_{ij}^n is the estimated time for a Bus on route b to travel from stop sequence i to j when the Bus is expected to start from stop sequence i in a time slot n. For each route b, the algorithm hence requires to maintain a small database of T_{ij}^n having $(M-1)*N$ entries for $i\in\{1,2,\cdots M-1\}$ and $n\in\{0,1,\cdots,N-1\}$.

We now present the proposed algorithm to initiate and update the T_{ij}^n for a particular route b. Initial values of T_{ij}^n are mean of the time taken by any bus on the route b to travel from stop sequence i to stop sequence j. Note that in the subsequent part of this paper, we drop the bus index b for brevity and we refer *bus stop sequence* as *bus stop*. Let, t_i^a be the actual arrival time at a bus stop i, then the time τ_{ij} required to travel from stop i to j, when it started from stop i in a time index n, becomes

$$\tau_{ij}^n = t_j^a - t_i^a. \tag{1}$$

When the τ_{ij}^n is known, the time estimate T_{ij}^n is proposed to be updated using the *update equation* (2).

$$T_{ij}^n = \alpha T_{ij}^n + (1-\alpha)\tau_{ij}^n \qquad \forall n \tag{2}$$

where $\alpha<1$ is the weightage given to the previous arrival time estimates and $(1-\alpha)$ is the weightage associated with the current travel time. If α is high then more weightage is given to history than the current dynamic traffic conditions and vice versa. The equation (2) has significant implication than that appears. Once a

Bus travels through a particular path the estimated travel time is updated not only based on the past statistics but also based on the current events, weather, road conditions and all other parameters that defines the speed of any vehicle. We thus get more accurate estimate of the bus arrival time on a particular stop by considering statistics as well as the dynamic conditions. Ideally, for each bus route b, as soon as new t_u^a is received we can compute corresponding τ_{ij}^n and hence corresponding T_{ij}^n can be be updated. However, based on the dynamic traffic pattern and computational power, the update of T_{ij}^n with different schedules, like hourly, daily or weekly, can be considered.

The database of T_{ij}^n contains $N * (M - 1)$ entries of T_{ij}^n for each Bus route in a city. If there are around 300 routes in a city with an average of $M = 30$ stops for each route, then by considering $N = 24$ time slots in a day and using 2 bytes to store a time duration index, the size of $T_(ij)^n$ becomes only $432KB$ or $0.43MB$. This tiny database can also be easily downloaded along with mobile application and can be further updated frequently by periodic contact to server.

Once the T_{ij}^n database has been updated, the estimated time of arrival at stop v (t_v^e), when the bus was last spotted at stop u at time t_u^a, can be computed using the following recursive algorithm.

$$
\begin{aligned}
t_{u+1}^e &= t_u^a + T_{u,u+1}^{fs(t_u^a)} \\
t_{u+k}^e &= t_{u+k-1}^e + T_{u+k-1,u+k}^{fs(t_{u+k-1}^e)} \\
&\quad for \quad k = 2, 3, \cdots, (v - u).
\end{aligned}
$$

Here, $fs(t)$ is a function which determines the time slot index n based on the input time parameter t. For instance, for $N = 24$, the time slot index $n \in \{0, 1, \cdots, N-1\}$ corresponds to the time slots $\{(00:00)-(00:59), (1:00)-(1:59), \cdots, (23:00)-(23:59)\}$, respectively. In this case if $t = 1:15$, then function $fs(t)$ results in time slot index $fs(t) = 1$. The estimated time t_v^e of arrival at stop v can hence be communicated to the individual who has queried for the same from the mobile or web application.

4 Experimental Results

In this section we first describe the experimental setup for human participatory sensing for the bus arrival time information system. The experiments were carried out in the city of Mumbai for a Bus route from Mulund Bus Stop to Agarkar Chowk Andheri (AS 422 UP) and for return route from Agarkar Chowk Andheri to Mulund Bus Stop (AS 422 Down). In each direction, the bus travels approximately 17 Km and covers 25 stops. In this experiment, our team members volunteered to capture the bus stop arrival time information using the proposed human participatory mobile phone application. Twenty number of trips were done on the selected route. Without loss of generality, this experiment has been carried out in the peak monsoon season, during the working days and for a few selected time slots.

We compute the error in travel time estimate between every two intermediate stops on the route and a corresponding root mean square error. In the following results, the root mean square error has been averaged for all the 25 stops on the entire route. For a particular time slot and a particular route, as and when the arrival time of the bus is update at stop j using the human participatory mobile phone application, the estimate T_{ij} of the time taken by a bus to travel from stop i to the adjacent stop j can be made using equation (2). For the route AS422-Down and the bus starting from the originating stop in a time slot 10am-11am, the root mean square error with respect to α has been shown in Fig. 5. In this case, the α that minimizes the RMSE can be selected to be $\alpha = 0.6$.

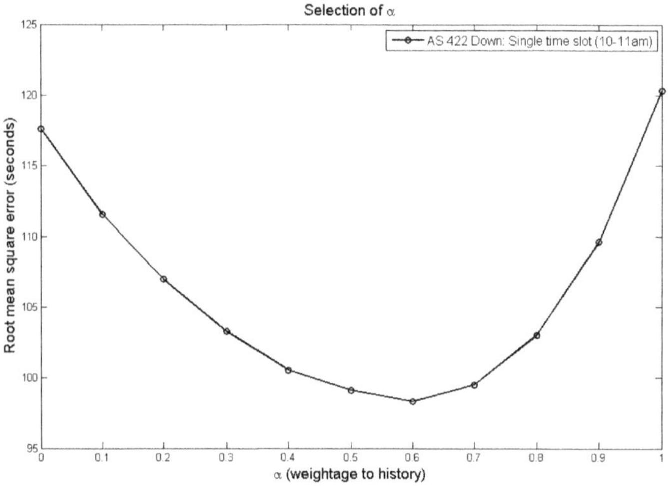

Fig. 5. Selection of α for a particular route and a time slot

For the same set-up, the root mean square error with respect to experiment numbers has been shown in Fig. 6. As observed, for a particular time slot with an increase in number of experiments, the RMSE presents a reducing trend.

The Fig. 7 shows the RMSE plots for considering following cases of bus route and time slots: (a) AS 422 Down for 10am to 11am (single time slot) (b) AS 422 Down for 10am to 11 am "off-peak" and 6pm to 9 pm "peak" (multiple time slots) (c) AS 422 UP for 4pm-8pm (multiple time slots). For this figure, we observe that the RMSE of a single time slot is significantly better than that of the multiple time slots. The RMSE further degrades by considering both peak and off peak scenarios. This again motivates the importance of the proposed algorithm which is based on updating the estimates for the smaller segments of the time slots, independently.

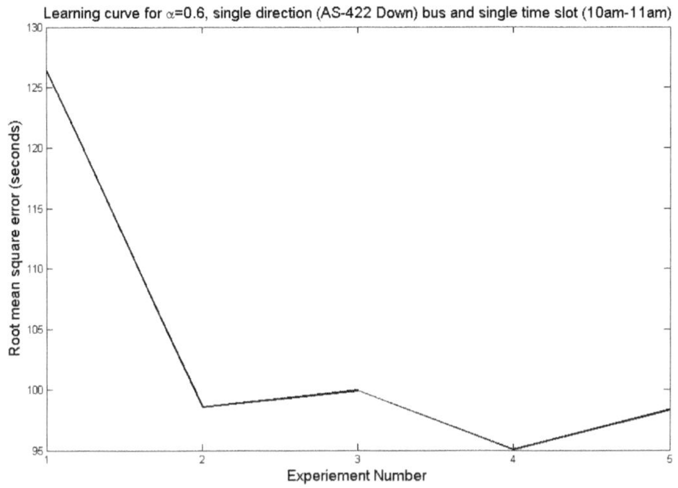

Fig. 6. Learning for $\alpha = 0.6$

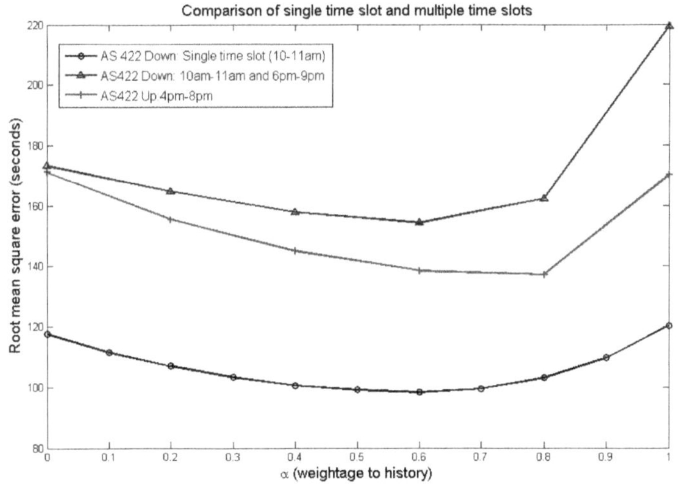

Fig. 7. Root mean square error for different routes and various time slot with respect to α

5 Conclusion and Future Work

We have developed a human participatory mobile phone application using which a passenger can update the instantaneous bus stop arrival time information while traveling. A novel *arrival time estimation algorithm*, based on the slotted time segments during a day, has been proposed in this paper. In order to make the traveling in public transport more comfortable, the proposed system provides the time estimates at the queried stop either on the mobile phone or on the console. The performance of the proposed system was analyzed from the experiment carried out for the public transport system in the city of Mumbai.

The future work on this research will include the data aggregation and distributed processing of participatory sensing data from more than one users. We also envision to develop a system which uses the GPS coordinates from the available mobile phones in an energy efficient manner, opportunistically.

Acknowledgments. We sincerely thank Mr. Anil Patil for his support for the graphics in the paper.

References

1. Lin, W.H., Zeng, J.: Experimental study of real-time bus arrival time prediction with gps data. Journal of the Transportation Research Record 1666, 101–109 (1999)
2. Shalaby, A.: Prediction model of bus arrival and departure times using AVL and APC data. Journal of Pub. Transp., 41–61 (2004)
3. Padmanaban, R.P.S., Vanajakshi, L., Subramanian, S.C.: Estimation of bus travel time incorporating dwell time for apts applications. In: IEEE Intelligent Vehicles Symposium, pp. 955–959 (2009)
4. Thiagarajan, A., Biagioni, J., Gerlich, T., Eriksson, J.: Cooperative transit tracking using smart-phones. In: The 8th ACM Conference on Embedded Networked Sensor Systems, SenSys 2010, pp. 85–98. ACM, New York (2010)
5. Hatem, B.A., Habib, H.: Bus management system using rfid in wsn. In: European and Mediterranean Conference on Information Systems, EMCIS 2010 (2010)
6. Menezes, B., Laddhad, K., Karthik, B., Dutta, K.: Challenges in rfid deployment a case study in public transportation (2006)
7. Gammer, N.: An appraisal of QR code use to deliver bus arrival time information at bus stops in Southampton. Master's thesis, MSc (Transportation Planning and Engineering), University of Southamptom, Faculty of Engineering and Environment, Transportation Planning and Engineering (December 2011)
8. Kuczynski, J.: QR code travel information service launched in Tokyo. Wireless World Japan (August 2006), `http://w2japan.blogspot.in/2006/08/qr-code-travel-information-service.html` (last accessed: February 21, 2012)
9. Burke, J., Estrin, D., Hansen, M., Parker, A., Ramanathan, N., Reddy, S., Srivastava, M.B.: Participatory sensing. In: Workshop on World-Sensor-Web (WSW 2006): Mobile Device Centric Sensor Networks and Applications, pp. 117–134 (2006)
10. Bulusu, N., Chou, C.T., Kanhere, S., Dong, Y., Sehgal, S., Sullivan, D., Blazeski, L.: Participatory sensing in commerce: Using mobile camera phones to track market price dispersion
11. Campbell, A.T., Eisenman, S.B., Lane, N.D., Miluzzo, E., Peterson, R.A.: People-centric urban sensing. In: The Second Annual International Wireless Internet Conference (WICON), pp. 2–5. IEEE Computer Society Press (2006)
12. Lane, N.D., Eisenman, S.B., Musolesi, M., Miluzzo, E., Campbell, A.T.: Urban sensing systems: Opportunistic or participatory. In: Proc. ACM 9th Workshop on Mobile Computing Systems and Applications, HOTMOBILE 2008 (2008)

13. Rekimoto, J.: Sensonomy: intelligence penetrating into the real space. In: Proceedings of the 14th International Conference on Intelligent User Interfaces, IUI 2009, pp. 3–4. ACM, New York (2009)
14. Dua, A., Bulusu, N., Feng, W.-C., Hu, W.: Towards trustworthy participatory sensing. In: Proceedings of the 4th USENIX Conference on Hot Topics in Security, HotSec 2009, p. 8. USENIX Association, Berkeley (2009)
15. Reddy, S., Samanta, V., Burke, J., Estrin, D., Hansen, M., Srivastava, M.: Mobisense mobile network services for coordinated participatory sensing. In: International Symposium on Autonomous Decentralized Systems, ISADS 2009, pp. 1–6 (2009)

Scheduling of Vehicles in Transportation Networks

Dariusz Kowalski[1], Zeev Nutov[2], and Michael Segal[3,*]

[1] Department of Computer Science, University of Liverpool, Liverpool, UK
[2] Department of Computer Science, The Open University of Israel, Raanana, Israel
[3] Department of Communication Systems Engineering, Ben-Gurion University of the Negev, Beer-Sheva, Israel

Abstract. In this paper we consider online vehicle scheduling problems for different network topologies under various objective functions: minimizing the maximum completion time, minimizing the largest delay, and minimizing the sum of completion times and present a number of provable approximate solutions.

1 Introduction

Classical traffic scheduling and control theory has been recently compromised by new challenges arising from green transportation, vehicular networks and automated vehicles. This paper studies vehicle scheduling algorithms in different types of networks motivated by scheduling strategies in various types of underlying network topologies and under different measures of performances, optimizing global and local latencies, as well as simplified measure of energy consumption (or CO2 emission). A network is represented as an undirected graph $G = (V, E)$. An (undirected) edge $(u, v) \in E$ between two nodes $u, v \in V$ represents a unit segment (of road) between two locations u and v. A set $C = \{c_1, \ldots, c_k\}$ of k vehicles should be routed through the network, where each vehicle has its own source s_i and destination node d_i. The capacity of each node (in terms of the number of vehicles it can keep at a time) is unlimited; however, each time only one vehicle can pass an edge in either direction but not in both. We consider scenarios in which all vehicles start their routes at the same time. We define a completion time of a vehicle as the time measured since the beginning till the vehicle reaches its destination. Let a delay of a vehicle be defined as a difference between the completion time of the vehicle and the length of its route (i.e., the number of edges in its route). Our goal is to move all the vehicles towards their destinations along some shortest path from its source to its destination (either given or to be chosen by the algorithm) under one or more of the following objectives:

* The work by Michael Segal has been partly supported by France Telecom, European project FLAVIA and Israeli Ministry of Industry, Trade and Labor (consortium CORNET). Email: segal@cse.bgu.ac.il

A. Vinel et al. (Eds.): Nets4Cars/Nets4Trains 2012, LNCS 7266, pp. 124–136, 2012.

- (*MinMakespan*) Minimizing the maximum completion time, i.e., the time when the last vehicle reaches its destination.
- (*MinMax*) Minimizing the largest delay of the vehicles.
- (*MinSum*) Minimizing the total sum of completion times over all vehicles.

The length of the scheduling path produced by some particular vehicle is the number of edges in this path. Observe that the optimum algorithm for the MinSum objective function is also the optimum solution if we replace "completion times" by "delays" in the specification of the MinSum objective function; this is because the sum of completion times is equal to the sum of delays plus the sum of the lengths of the routes, and the latter is fixed and straightforward to compute for a given input. We consider different graph topologies: grid, rooted tree and bipartite thrackle (without shared endpoints). In addition, we distinguish between the settings where, for every vehicle, the path from its source node to its destination node is given or not; in both cases, as we noted earlier, the routes must be shortest possible between the sources and destinations.

2 Related Work

In [14], Leighton et al. showed that for any network and any set of packets whose paths through the network are fixed and edge-simple, there exists a schedule for routing the packets to their destinations in $O(c + d)$ steps using constant-size queues, where c is the congestion of the paths in the network, and d is the length of the longest path. Mansour and Patt-Shamir [19] and also Cidon et al. [7] showed that if packets are routed greedily on shortest paths, then all of the k packets reach their destinations within $d + k$ steps. These schedules may be much longer than optimal, however, because k may be much larger than c. Rabani and Tardos [22], and Ostrovsky and Rabani [20] extended the main ideas used in [14] and in the centralized algorithm presented in [16] to obtain online local control algorithms for the general packet routing problem that produce near-optimal schedules. Some related result for trees can be found in [6]. Liu and Zaks [18] studied the greedy algorithm for delivering messages with deadlines in synchronous networks. They considered bottleneck-free networks, in which the capacity of each edge leaving any processor is at least the sum of the capacities of the edges entering it. For such networks where there is at most one simple path connecting any pair of vertices, they have shown a necessary and sufficient condition for the initial configuration to have a feasible schedule, and proved that if this condition holds then the greedy algorithm, that chooses at each step the most urgent messages (those with closest deadlines), determines such a feasible schedule. Adler et al. [1] dealt with the time-constrained packet routing problem when one wants to schedule a set of packets to be transmitted through a multinode network, where every packet has a source and a destination (as in traditional packet routing problems) as well as a release time and a deadline. The objective is to schedule the maximum number of packets subject to deadline constraints. The problem is known to be NP-complete even when the underlying topology is a linear line [2]. For the buffered case, [1] provides logarithmic factor approximation algorithms for the time-constrained scheduling problem with weighted packets on trees and meshes. Leung et al. [17] considered a problem of routing unit-length, -time messages in different types of networks under various restrictions of four parameters: source node, destination node, release time, and deadline. Peis et al. [21] considered different routing problems for fixed and variable paths for the grid topology. In [13], a line topology has been considered in the setting similar to this work: for directed lines optimal local algorithms have been designed, while for undirected lines no local algorithm is optimal, for each of the three objective functions. Instead, 2-approximation local solutions have been proposed for the latter setting, based on the optimal solutions for directed lines.

Many algorithms designing efficient gathering scheme in the context of wireless networks have been considered before, see e.g. [3–5, 11, 23]. The aim is to minimize the number of steps (*makespan*) needed to send all messages to the base station. Opposite to our model, a node cannot both receive and transmit simultaneously. Moreover, during a step only non interfering transmissions can be done.

This paper is organized as follows. First we consider grid network topology. Next, at Section 4, we turn our attention to the rooted tree networks. Finally, we consider bipartite thrackle networks. We conclude the paper at Section 6.

3 Grid Networks

We start by considering the grid version of the problem. We are given a unit-sized grid of size $n \times n$. We assume that all of the paths for the routing vehicles are given, the vehicles are located in the bottom left quarter of the grid and all the paths are dimension order paths (as in [1]). Dimension order routing, which is commonly used on the mesh, requires that each vehicle travels along its source row to the correct column and then along that column to the correct row, or vise versa.

3.1 Optimizing MinMakespan Objective

First, we notice that if every two shortest paths have different origin and destination nodes, then the approximation ratio for MinMakespan will be 2, since each path of length l can have at most $l - 1$ intersections points with other paths. In the general grid problem, the total number of delays can be as large as $\Omega(n^2)$, and applying the longest-remaining-distance-first strategy, as was efficient for linear networks [13], might not work. In order to see this, suppose that we have $k = n/2$ vehicles starting at the same time and having the same distance to go for their destinations, as deployed at Figure 1. When two of them meet then one of them is delayed by 1. Then the same vehicle that caused this delay (by itself it is not delayed) meets another vehicle, which becomes delayed by 1, and so on. At the end of this part, there is one non-delayed vehicle and $n/2 - 1$ vehicles delayed by 1. If we apply the same procedure to these $n/2 - 1$ vehicles we get

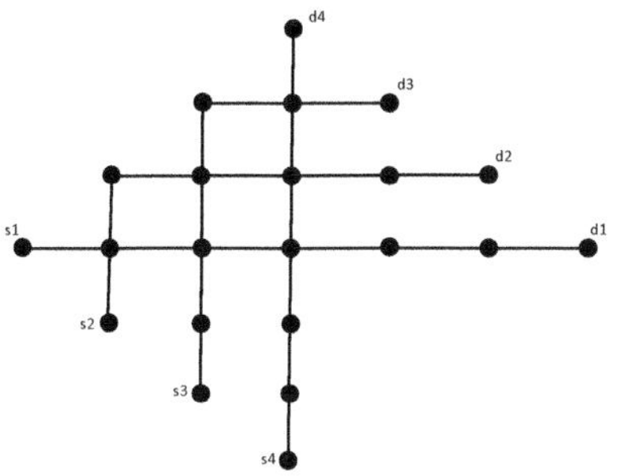

Fig. 1. A scenario producing many delays in a grid when longest-remaining-distance-first strategy is implemented. The number of vehicles/paths is $n/2$.

$n/2-2$ vehicles delayed by 2 and so on. Thus, at the end of this algorithm some of the vehicles may suffer a linear (in n) time delay. On the other hand, the optimum solution for the example in Figure 1 has MinMakespan of at most the distance-to-go plus one, as this makespan can be obtained for example when in case of collision a priority is always given to a vehicle going up in the grid. In fact, Figure 1 provides a lower bound of 1.5 approximation ratio for *any* greedy strategy that can be applied in order ro resolve collisions between the vehicles, since the greedy strategy will produce the solution of at least the distance-to-go plus the distance-to-go divided by 2.

Next we argue that any algorithm (resolving a contention at node in an arbitrary way) achieves $o(n)$ approximation ratio. In order to see this, we first argue that if every path has at most $o(n)$ intersections we are done. Otherwise, assume that some path has $\Omega(n)$ intersections. If the length of this path is $\omega(1)$, we are again done. Otherwise assume that the length of this path is constant. If the length of some other path is $\omega(1)$, we are done (since the approximation factor will be $n/\omega(1) = o(n)$). Otherwise, the length of all paths is constant. Hence, we have a situation with all path lengths being constant (possibly different for different paths) and some path has $\Omega(n)$ intersections. In this case, the optimum solution schedule takes $\Omega(n)$ time as any algorithm.

Centralized Greedy Longest-to-Destination algorithm with stages. In order to achieve better approximation factor we suggest another approach. Recall that all our paths are dimension order paths. Let p be the upper bound on the number of segments in each path (in our case $p=2$). So, we first solve the problem for vehicles going up (we freeze all the vehicles that start horizontally). We can do it optimally for MinMakespan objective according to the following. We adopt the smallest slack time algorithm from [17], designed in a slightly different setting with deadlines. In this model, the goal is to maximize the number of vehicles that arrive to their destinations within their deadlines. The algorithm in [17] gives priority to the vehicle with the smallest maximum time span that this vehicle can be delayed by without missing its deadline (i.e., that minimizes time to the deadline minus the remaining length of the route). Since the deadlines in this setting could be chosen arbitrarily large for all vehicles, this is equivalent to giving priority to the vehicle whose destination node is currently farthest away in our model; i.e., both executions — the one in the model with deadlines and the other in our model — result in the same vehicle schedules. Leung et al. [17] proved that this algorithm is optimal for the case of MinMakespan objective criteria for directed graph, in the model with deadlines. By equivalency of executions, the corresponding Furthest-from-Destination algorithm is optimal in our model when all vehicles travel in the same direction.

After all such vehicles arrived at their node of changing direction to the horizontal one, we run the second stage in which the optimum algorithm is applied only for horizontal vehicles (prioritizing the vehicle having largest remaining *horizontal* distance). We continue this procedure until the last segment; in total, we have at most $p+1$ stages.

We argue that the above algorithm achieves $p+1$ approximation. Suppose we are given a set of paths $\mathcal{P}$, each of at most p segments. Consider stage i of the algorithm, for $1 \le i \le p+1$. Consider a line L, either horizontal or vertical, in a grid such that L contains at least one segment of some path in $\mathcal{P}$ which is considered in stage i (i.e., either i-th segment, if the path started with vertical segment, or $(i+1)$-th segment otherwise). Observe that during stage i propagation of vehicles along different parallel lines L are independent, in the sense that they do not collide with each other. Let $\mathcal{L} = \mathcal{P}|_L$ denote the family of sets $P \cap L$, over all $P \in \mathcal{P}$. Observe that $\mathcal{L}$ contains one-directional sub-paths of L. Let OPT($\mathcal{P}$) denote the optimum solution for $\mathcal{P}$, and OPT($\mathcal{L}$) be the optimum solution for input $\mathcal{L}$. By $|\text{OPT}(\mathcal{P})|$ we denote the makespan of OPT($\mathcal{P}$), and by $|\text{OPT}(\mathcal{L})|$ we denote the makespan of OPT($\mathcal{L}$). Note that $|\text{OPT}(\mathcal{L})|$ is at most $|\text{OPT}(\mathcal{P})|$. Indeed, we could consider OPT($\mathcal{P}$) solution in the time period between the first vehicle starts to traverse its segment in $\mathcal{L}$ (i.e., arrives at the beginning of its segment) till the last vehicle finishes its segment in $\mathcal{L}$ (i.e., arrives at the end point of its segment in $\mathcal{L}$), with respect only to the vehicles traversing segments in $\mathcal{L}$. Since we have to show the algorithm on the input set $\mathcal{L}$, we need to modify the above execution as follows. If some of the vehicles has not yet arrived till the starting point of its segment in $\mathcal{L}$ during the considered period of OPT($\mathcal{P}$), in the corresponding algorithm for input $\mathcal{L}$ we assume that it stays idle in its starting point; similarly, if a vehicle leaves its segment in $\mathcal{L}$ during the considered part of the execution of OPT($\mathcal{P}$), in the corresponding algorithm for $\mathcal{L}$ we assume that it stays in its destination point. Clearly, the length of this time period is at most the length of the whole execution of OPT($\mathcal{P}$), the modified algorithm for the purpose of input $\mathcal{L}$ is of the same length and it constitutes a feasible solution of vehicle scheduling for $\mathcal{L}$. This concludes the proof that each stage is of length at most $|\text{OPT}(\mathcal{P})|$. Finally, the analysis is concluded by recalling the upper bound $p+1$ on the number of stages.

Local algorithm. In each node, the decision which vehicle should be pushed to the next node in its path is taken according to the lexicographic order on pairs of parameters $(p+1-stage, segm\_dist)$ defined for vehicles residing at this node, where *stage* denotes the stage number of the vehicle and *segm_dist* stands for the remaining distance to travel by the vehicle in the current segment. The smallest vehicle according to this order is pushed forward to its next node. This algorithm is clearly local, i.e., a decision about pushing a vehicle is made based on the information carried by vehicles located at the node. Moreover, its Makespan is not smaller than in the previous centralized algorithm, i.e., it is bigger than the minimum Makespan by factor at most $p + 1$. To see this, it is enough to observe, using a straightforward inductive argument on the number of rounds, that at any time each vehicle is not further from its final destination than during the centralized scheduling described before.

3.2 Optimizing MinSum and MinMax Objectives

In fact, any algorithm (resolving a contention at node in arbitrary way) provides a $O(\sqrt{k})$-approximation for the MinSum criteria for grid structure. For this, suppose that the optimal algorithm (minimizing the sum of delays) schedules all the vehicles to their destinations at maximal arrival time T. Let us look at some vehicle c' and estimate how many vehicles can delay c' on its way to the destination. Knowing that the maximal arrival time in optimal solution is T, we can conclude that the distance of any vehicle to its destination is at most T and the total number of vehicles that can be located at the same source is also T. Consequently, the maximal number of vehicles that can delay vehicle c' is at most $O(T^2)$. At the other hand this number can not be more than k. Taking $T^2 = k$, we obtain the required approximation bound. We note here, that this holds for any bounded degree graph as well.

In order to deal with MinMax objective function, we first solve the decision problem by looking at all the possible configurations of vehicles that allow all of them to arrive to their destinations under the constant maximum delay δ. Let us consider the example in Figure 2.

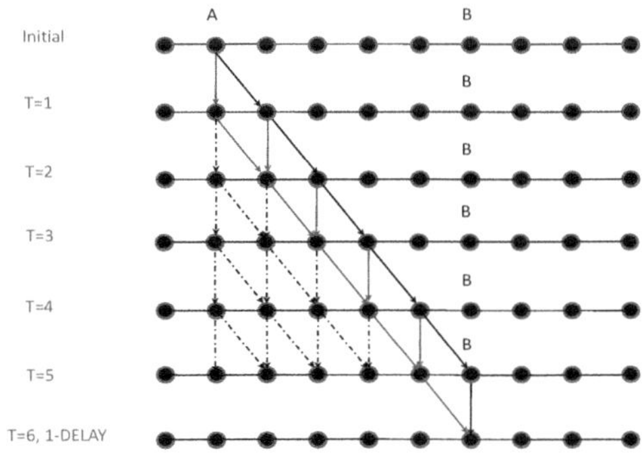

Fig. 2. Possible configurations for the movement of the vehicle from source A to destination B

We explain our idea on linear network, though it works for any undirected graph with n nodes. Every linear network depicted in Figure 2 represents a possible situation in some particular moment of time. Initially some vehicle is located at point A and wants to go to point B. At the next moment ($T = 1$) the vehicle can still be delayed at A or moved one step towards B. This is shown in a horizontal linear network at time $T = 1$. All consequent networks show possible locations of the vehicle at every consequent moment of time. We also draw the directed arrows from the current possible location of the vehicle to the next possible location of the vehicle in the consequent moment of time. Since the distance between A and B is 5, the length of the directed path that shows the possible locations of the vehicle that was not suffering any delay should be 5 — this is depicted in Figure 2 as the directed black solid path. If we would allow this vehicle to suffer up to one delay, then we should also consider the paths defined by the directed red edges (in addition to the directed solid black edges). In such a way we can build, for each vehicle, all relevant edges that it can contribute to the obtained graph under the constraint that each vehicle can suffer up to δ delays. The question we are asking now is whether there are k vertex-disjoint paths (we allow situations when the endpoint of one path coincides with some node of another path) between k pairs (s_i, d_i) of sources and destinations in the obtained directed acyclic graph?

There are several papers that studied the problem of finding k vertex-disjoint paths in graphs. For general undirected graphs, the vertex-disjoint paths problem is solvable in polynomial time for any fixed k [24] and is NP-complete if k is a part of the input [12] (here we mention that the problem can be 2-approximated [15] for minimizing the sum of the length of paths for $k = 2$, which can be used for solving our scheduling problem for MinSum objective function). For general directed graphs, the problem is NP-complete even when $k = 2$ [10]. For a directed acyclic graph and constant k, Li et al. [15] gave two efficient pseudo-polynomial time algorithms and Fleischer et al. [9] gave an efficient fully polynomial-time approximation scheme. The currently best solution is due to Yu et al. [25] who actually solve the more general problem of finding k pairwise vertex-disjoint paths such that the maximum length of these k paths is minimized. The optimal algorithm requires $O(n^{k+1}M^{k-1})$ time, and the presented approximation scheme requires $O((1/\varepsilon)^{k-1}n^{2k}\log^{k-1} M)$ time, where ε is the given approximation parameter and M is the length of the longest path in an optimal solution. Thus in order to obtain a solution to our problem we can use the algorithms in [25] in order to obtain a solution to the decision problem with a subsequent binary search scheme. Below we consider the problem of finding the maximum number of pairs from K via node-disjoint paths. We give a $\sqrt{n}$-approximation algorithm for this problem, where $n = |V|$, and show that the problem admits no $O\left(n^{1/2-\varepsilon}\right)$-approximation, unless P=NP.

In other words, the goal is to find a maximum size set of pairs

$$K' = \{(s_1, d_1), \ldots, (s_{k'}, d_{k'})\} \subseteq K$$

and a collection $\{P_1, \ldots, P_{k'}\}$ of pairwise node-disjoint paths in G, such that each P_i is an $s_i d_i$-path, i.e., a path from s_i to d_i. In the other version of the problem, the paths are required to be only pairwise internally-disjoint, so the end node of one path can be located in another path. We call this version the Internally-Disjoint Paths problem.

Theorem 1. Node-Disjoint Paths *and* Internally-Disjoint Paths *admit a* $\sqrt{n}$*-approximation algorithm.*

Proof. Given a graph $G = (V, E)$ and $K \subseteq V \times V$, let $\Pi_G(K)$ be a set of paths obtained by picking the shortest sd-path in G for every $(s, d) \in K$. The algorithm is a standard greedy algorithm, parameterized by ℓ, eventually set to $\ell = \sqrt{n}$.

Algorithm for Node-Disjoint Paths
Initialization: $K' \leftarrow \emptyset$, $\mathcal{P} \leftarrow \emptyset$.
While there is $P \in \Pi_G(K)$ with at most ℓ nodes do:
  Let $(s, d) \in K$ be the pair that P connects.
  Add P to $\mathcal{P}$, move (s, d) from K to K',
  and remove P from G.
EndWhile
If there is $P \in \Pi_G(K)$, add P to $\mathcal{P}$ and add to K' the pair $(s, d) \in K$ that P connects.

It is clear that the algorithm runs in polynomial time and computes a feasible solution to the problem. Let $\mathcal{P}'$ be the set of paths in $\mathcal{P}$ added during the while-loop, and note that every path in $\mathcal{P}'$ has at most ℓ nodes. Let opt be the optimal solution value to the problem, and let opt$'$ be the optimal solution to the residual problem after the end of the while-loop. Clearly, $\mathsf{opt}' \leq n/\ell$. Whenever a path P is added to $\mathcal{P}$, the optimum solution value has decreased by at most the number of nodes in P. Hence $\mathsf{opt} - \mathsf{opt}' \leq \ell \cdot |\mathcal{P}'|$. We claim that $|\mathcal{P}| \geq \mathsf{opt}/\ell$ for $\ell \geq \sqrt{n}$, hence Theorem 1 follows by substituting $\ell = \sqrt{n}$. To see this, consider two cases. If $\mathsf{opt}' = 0$ then $|\mathcal{P}| = |\mathcal{P}'| \geq \mathsf{opt}/\ell$. If $\mathsf{opt}' \geq 1$ then $|\mathcal{P}| = |\mathcal{P}'| + 1 \geq \frac{\mathsf{opt}}{\ell} - \frac{\mathsf{opt}'}{\ell} + 1 \geq \frac{\mathsf{opt}}{\ell} - \frac{n}{\ell^2} + 1$. In both cases, we have $|\mathcal{P}| \geq \mathsf{opt}/\ell$ for $\ell \geq \sqrt{n}$, as claimed.

The algorithm for Internally-Disjoint Paths is similar, except that a path P is added to the partial solution $\mathcal{P}$ if it has at most ℓ internal nodes, and then we remove only the internal nodes of P from G. The proof of the approximation ratio is identical to the one of the Node-Disjoint Paths case. The theorem follows.

Theorem 2. *For any* $\varepsilon > 0$, Node-Disjoint Paths *and* Internally-Disjoint Paths *admit no* $O\left(n^{1/2-\varepsilon}\right)$*-approx-imation algorithm, unless P=NP.*

To prove Theorem 2, we reduce the problem to the Edge-Disjoint Paths problem, which has approximation threshold $|E|^{1/2-\varepsilon}$ for instances with $|E| \geq |V|$. This is done by a standard reduction that converts node-disjoint paths into edge-disjoint paths.

Given an instance $< G = (V, E), K >$ of Edge-Disjoint Paths, define an instance $< G' = (V', E'), K' >$ of Node-Disjoint Paths as follows. The graph G'

is obtained from G by replacing every node $v \in V$ by the two nodes v^{in}, v^{out} and the edge (v^{in}, v^{out}), and then replacing every edge $(u,v) \in E$ by the edge (u^{out}, v^{in}). The set of pairs K' is defined by $K' = \{(s^{in}, d^{out}) : (s,d) \in K\}$. Every solution to the Edge-Disjoint Paths problem for input $< G, K >$ corresponds to some solution to the Node-Disjoint Paths problem for the corresponding input $< G', K' >$ of the same size, and vice versa. Moreover, given one of the solutions the other can be computed in polynomial time.

Clearly, $|E'| = |E| + |V|$. Now suppose to the contrary that there exists an approximation algorithm for the Edge-Disjoint Paths problem with ratio $|E'|^{1/2-\varepsilon}$, which by definition must hold for every input $< G', K' >$. This implies that there exists an approximation algorithm for the the Node-Disjoint Paths problem with ratio $(|E| + |V|)^{1/2-\varepsilon} \leq (2|E|)^{1/2-\varepsilon} \leq \sqrt{2}|E|^{1/2-\varepsilon}$, by considering the correspondence between inputs $< G', K' >$ and $< G, K >$ for the Edge-Disjoint Paths and the Node-Disjoint Paths problems, respectively. For $|E|$ large enough, this implies approximation $|E|^{1/2-\delta}$ for some $\delta > 0$. This contradicts the approximation threshold for the Edge-Disjoint Paths problem.

At this stage we are able to provide an $O(\sqrt{n})$ approximation to the number of vehicles that arrive to their destination and suffer no more than δ delay. The question is how the delay will change if we allow all the vehicles to reach their destinations. For this, we apply $O(\sqrt{n})$ approximation algorithm explained above a number of times. After the first time we obtain a number of paths for vehicles that allow them to reach their destinations with δ delay. Call this paths 1st *schedule group*. Next, we remove all of these paths, and repeat the algorithm by finding the paths obtained at second run (2nd *schedule group*). We continue this procedure until no paths for vehicles can be produced and all the vehicles arrived to their destination. Notice, that this process can be repeated no more than $\sqrt{n}\log n$ times. It means that we obtained $\sqrt{n}\log n$ schedule groups. In order to schedule all the vehicles, we divide the time slots into $\sqrt{n}\log n$ sets corresponding to $\sqrt{n}\log n$ schedule groups. The ith schedule group receives $(i \mod \sqrt{n}\log n)^{th}$ time slot to schedule the vehicles on their paths. In such way, we guarantee that all vehicles arrive at their destinations after each vehicle experiences at most $\delta\sqrt{n}\log n$ delays.

We mention here that in order to solve the problem under MinMax delay performance measure, we can follow the same approach as presented above for linear networks, by obtaining 3-dimensional directed acyclic graph and running algorithm for finding k vertex-disjoint paths between sources and destinations.

4 Rooted Trees

In the rooted tree version problem, the vehicles should go from theirs sources to their destinations by their shortest paths. Since we have a tree, every path is defined uniquely, i.e., it can be either either vertically upward, vertically downward or upward-downward: go up to the least common ancestor of source and destination and then downward. For our problem on rooted tree we can use approach similar to the one for grids: we first push only vehicles going up until all of them arrive to their highest point and then looking at the vehicles only going down. In such a way we achieve 3 approximation for this case for MinMakespan criterion.

In order to show the approximation ratio for MinSum criteria we apply shortest remaining distance first algorithm. In this algorithm we always give a priority to the vehicle with the shortest remaining distance to go, where ties are broken arbitrary. For this, suppose that the optimal algorithm (minimizing the sum of delays) schedules all the vehicles to their destinations at maximal arrival time T. Let us look at some vehicle c' and ask a question about how many vehicles can delay c' on its way to destination. Knowing that the maximal arrival time in optimal solution is T, we can conclude that the distance of any vehicle to its destination is at most T and therefore, there are at most $3T$ vehicles that can delay c' on every path's segment (upward or downward) (T vehicles from behind and $2T$ vehicles ahead of c'). We notice that, in general, some vehicle c'' following the same direction as c' can delay c' more than once; however, in this case c'' should also be delayed by some other vehicle. Moreover, each such "repeated" delay can be charged to the vehicle from the set of $2T$ vehicles that will never delay c'. It follows that the total number of delays is at most $6T$ (3T downward and 3T upward), which leads to 7-approximation proof.

5 Bipartite Thrackle Network

A topological graph G is a graph drawn in the plane with vertices as points and edges as curves connecting its endpoints such that any two edges have at most one point in common. G is called a *thrackle* if every pair of edges intersects. Suppose, we are give a bipartite thrackle (see Figure 3).

In a bipartite thrackle every two lines are intersected in one point and no three lines have a joint point. We assume that passing each segment by vehicle in obtained bipartite thrackle graph takes one unit of time.

The vehicles are located at $A_1, \ldots, A_n$ points (as shown in Figure 3) and should arrive to $B_1, \ldots, B_n$ destinations. Thus, in our case $s_i = A_i$ and $d_i = B_i$. Let us consider the properties of the paths obtained by the following algorithm.

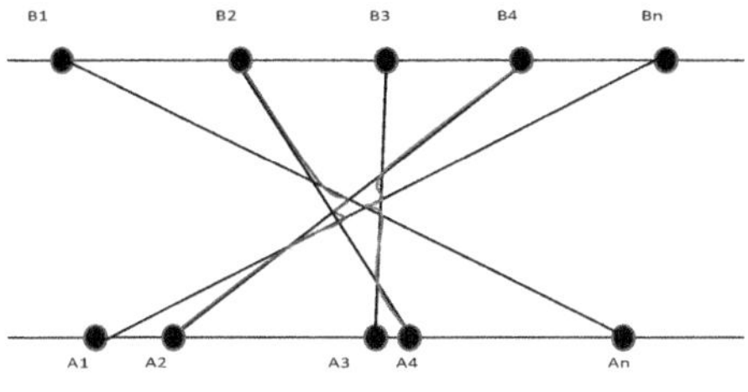

Fig. 3. Bipartite thrackle with switching paths

When a vehicle wants to move from A_i to B_i, we start at A_i and move up. Whenever we meet another intersecting line, the vehicle moves up to this line (as shown in Figure 3 by even numbered paths). Denote by l_i the length of such path between A_i and B_i in terms of number of segments in this path. First we will make few observations regarding the obtained paths.

1. Path that starts at A_i will surely end at B_i.
2. Every segment of each line belongs to only one path.
3. Adjacent paths (say, k^{th} and $(k+1)^{th}$) are touching by vertices, where k^{th} path is always to the left of $(k+1)^{th}$ path. Moreover, non-adjacent paths do not have joint points.

In fact, if we simply rotate Figure 3 by 90° degrees, we get an arrangement of n lines, where every pair intersects. The paths described above are simple the k-level paths in the arrangement – all points on the k-th path have exactly k-1 segments below them (that is, to their left). By Observation 3, if we take odd (or even) numbered paths, we would not have any intersections between them. Thus in order to schedule our vehicles to satisfy objective MinMakespan, we apply our algorithm separately for odd numbered vehicles, and then for even numbered vehicles.

It is well known that the upper bound for the length of the obtained paths (in terms of the number of segments) produced by our algorithm is $O(n^{\frac{4}{3}})$ [8].

Claim. On the other hand, at least one of the paths produced by the optimal solution algorithm satisfying MinMakespan criterion has at least length n.

Proof. Let us look at the middle path of order $m = \frac{n+1}{2}$ if n is odd, and of order $m = n/2$ if n is even. The path $A_{\frac{n}{2}}B_{\frac{n}{2}}$ partitions the area into 2 parts: each line $A_1B_n, A_2B_{n-1}, \ldots, A_nB_1$ starts in one part and ends at another part. It means that, the path $A_{\frac{n}{2}}B_{\frac{n}{2}}$ is intersected at least $n-1$ times.

Following the discussion above, we get that our algorithm achieves $O(n^{\frac{1}{3}})$ approximation for MinMakespan criterion.

Regarding the MinSum criteria, we note that by Observation 2 and Observation 3 we have $l_i \leq l_{i-1} + l_{i+1}$, where $2 \leq i \leq n-1$. On the other hand, it is clear the the best possible solution is achieved when every vehicle does not experience any delay and the total sum of completion times of all vehicles in this case is at least $\sum_{i=1}^{n} l_i$. If we apply the same algorithm as for the MinMakespan case, we obtain that the total sum of completion times of all vehicles is at most $3\sum_{i=1}^{n} l_i$. Approximation 3 for the sum of completion times follows.

6 Conclusions

At this paper we have considered vehicle scheduling problems for grid, rooted trees and bipartite thrackle networks under criteria of minimizing the maximum completion time, minimizing the largest delay, and minimizing the sum of completion times. We obtained polynomially solved approximate solutions for our problem. It would be interesting to generalize our scheme to more general networks' topologies.

References

1. Adler, M., Khanna, S., Rajaraman, R., Rosen, A.: Time-constrained scheduling of weighted packets on trees and meshes. Algorithmica 36, 123–152 (2003)
2. Adler, M., Rosenberg, A.L., Sitaraman, R.K., Unger, W.: Scheduling time-constrained communication in linear networks. In: Proc. 10th ACM Symp. on Parallel Algorithms and Architectures, pp. 269–278 (1998)
3. Bermond, J.-C., Nisse, N., Reyes, P., Rivano, H.: Minimum Delay Data Gathering in Radio Networks. In: Ruiz, P.M., Garcia-Luna-Aceves, J.J. (eds.) ADHOC-NOW 2009. LNCS, vol. 5793, pp. 69–82. Springer, Heidelberg (2009)
4. Bonifaci, V., Klasing, R., Korteweg, P., Stougie, L., Marchetti-Spaccamela, A.: Data Gathering in Wireless Networks. In: Graphs and Algorithms in Communication Networks. Springer, Heidelberg (2009)
5. Bonifaci, V., Korteweg, P., Marchetti-Spaccamela, A., Stougie, L.: An approximation algorithm for the wireless gathering problem. Oper. Res. Lett. 36(5), 605–608 (2008)
6. Busch, C., Magdon-Ismail, M., Mavronicolas, M., Wattenhofer, R.: Near-Optimal Hot-Potato Routing on Trees. In: Danelutto, M., Vanneschi, M., Laforenza, D. (eds.) Euro-Par 2004. LNCS, vol. 3149, pp. 820–827. Springer, Heidelberg (2004)
7. Cidon, I., Kutten, S., Mansour, Y., Peleg, D.: Greedy Packet Scheduling. SIAM J. Comput. 24(1), 148–157 (1995)
8. Dey, T.K.: Improved Bounds for Planar k-Sets and Related Problems. Discrete & Computational Geometry 19(3), 373–382 (1998)
9. Fleischer, R., Ge, Q., Li, J., Zhu, H.: Efficient Algorithms for k-Disjoint Paths Problems on DAGs. In: Kao, M.-Y., Li, X.-Y. (eds.) AAIM 2007. LNCS, vol. 4508, pp. 134–143. Springer, Heidelberg (2007)
10. Fortune, S., Hopcroft, J.E., Wyllie, J.: The directed subgraph homeomorphism problem. Theoret. Comput. Sci. 10, 111–121 (1980)
11. Gargano, L.: Time Optimal Gathering in Sensor Networks. In: Prencipe, G., Zaks, S. (eds.) SIROCCO 2007. LNCS, vol. 4474, pp. 7–10. Springer, Heidelberg (2007)
12. Karp, R.M.: On the computational complexity of combinatorial problems. Networks 5, 45–68 (1975)
13. Kowalski, D.R., Nussbaum, E., Segal, M., Milyeykovsky, V.: Scheduling Problems in Transportation Networks of Line Topology (2011) (manuscript), `http://www.cs.bgu.ac.il/~segal/sch.pdf`
14. Leighton, F.T., Maggs, B.M., Rao, S.B.: Packet routing and job-shop scheduling in O(congestion + dilation) steps. Combinatorica 14(2), 167–180 (1994)
15. Li, C.-L., McCormick, S.T., Simchi-Levi, D.: The complexity of finding two disjoint paths with min-max objective function. Discrete Appl. Math. 26(1), 105–115 (1990)
16. Leighton, F.T., Maggs, B.M., Richa, A.: Fast algorithms for finding O(Congestion + Dilation) packet routing schedules. Combinatorica 19(3), 375–401 (1999)
17. Leung, J.Y.-T., Tam, T., Young, G.: On-line routing of real-time messages. J. of Paral. and Dist. Computing 34, 211–217 (1996)
18. Liu, K.-S., Zaks, S.: Scheduling in synchronous networks and the greedy algorithm. Theor. Comput. Sci. 220(1), 157–183 (1999)
19. Mansour, Y., Patt-Shamir, B.: Greedy packet scheduling on shortest paths. Journal of Algorithms 14, 449–465 (1993)
20. Ostrovsky, R., Rabani, Y.: Universal $O(\text{congestion}+\text{dilation}+\log^{1+\varepsilon} N)$ local control packet switching algorithms. In: STOC, pp. 644–653 (1997)

21. Peis, B., Skutella, M., Wiese, A.: Packet Routing on the Grid. In: López-Ortiz, A. (ed.) LATIN 2010. LNCS, vol. 6034, pp. 120–130. Springer, Heidelberg (2010)
22. Rabani, Y., Tardos, E.: Distributed packet switching in arbitrary networks. In: STOC, pp. 366–375 (1996)
23. Revah, Y., Segal, M.: Improved algorithms for data-gathering time in sensor networks II: ring, tree, and grid Topologies. International Journal of Distributed Sensor Networks 5, 463–479 (2009)
24. Robertson, N., Seymour, P.D.: Graph minors. XIII. The disjoint paths problem. J. Combin. Theory Ser. B 63(1), 65–110 (1995)
25. Yu, C.-C., Lin, C.-H., Wang, B.-F.: Improved algorithms for finding length-bounded two vertex-disjoint paths in a planar graph and minmax k vertex-disjoint paths in a directed acyclic graph. Journal of Computer and System Sciences 76, 697–708 (2010)

Design and Prototype of a Train-to-Wayside Communication Architecture

Johan Bergs[1], Erwin Van de Velde[1], Daan Pareit[2], Dries Naudts[2], Milos Rovcanin[2], Ivan De Baere[3], Walter Van Brussel[4], Chris Blondia[1], Ingrid Moerman[2], and Piet Demeester[2]

[1] IBBT - PATS, Dept. of Mathematics and Computer Science, University of Antwerp
Middelheimlaan 1, 2020 Antwerpen, Belgium
{johan.bergs,erwin.vandevelde,chris.blondia}@ua.ac.be
[2] IBBT - IBCN, Dept. of Information Technology (INTEC), Ghent University
Gaston Crommenlaan 8 bus 201, 9050 Gent, Belgium
[3] Newtec Cy N.V., Laarstraat 5, 9100 Sint-Niklaas, Belgium
[4] Nokia Siemens Networks, Atealaan 34, 2200 Herentals, Belgium

Abstract. Telecommunication has become very important in modern society and seems to be almost omnipresent, making daily life easier, more pleasant and connecting people everywhere. It does not only connect people, but also machines, enhancing the efficiency of automated tasks and monitoring automated processes. In this context the IBBT (Interdisciplinary Institute for BroadBand Technology) project TRACK (TRain Applications over an advanced Communication networK), sets the definition and prototyping of an end-to-end train-to-wayside communication architecture as one of the main research goals. The architecture provides networking capabilities for train monitoring, personnel applications and passenger Internet services. In the context of the project a prototype framework was developed to give a complete functioning demonstrator. Every aspect: tunneling and mobility, performance enhancements, and priority and quality of service were taken into consideration. In contrast to other research in this area, which has given mostly high-level overviews, TRACK resulted in a detailed architecture with all different elements present.

1 Introduction

In the IBBT (Interdisciplinary Institute for BroadBand Technology) project TRACK (TRain Applications over an advanced Communication networK), one of the main research topics was the definition and prototyping of an end-to-end train-to-wayside communication architecture, which is also a research topic of the ESA ARTES project [17]. It is important to use a centralized approach for the communication system, connecting the onboard network transparently to the wayside, because it offers a more flexible and scalable solution. It allows for better coverage on board and joint bandwidth optimizations, traffic conditioning and differentiation.

A. Vinel et al. (Eds.): Nets4Cars/Nets4Trains 2012, LNCS 7266, pp. 137–150, 2012.

The TRACK communication architecture can be used to bring passenger Internet on the train, but more importantly, to enable a much broader spectrum of railway applications, such as CCTV (Closed Circuit TV) surveillance, train control and diagnostics, etc. If these kind of operational services are to be supported alongside passenger entertainment, a mechanism must be deployed to provide sufficient Quality of Service (QoS) guarantees in the dynamic train environment. The main topics that were researched in order to tackle these challenges, are:

- Analyze the behavior of relevant wireless technologies (satellite, Wi-Fi, 3G) in the dynamic train environment in terms of variations in bandwidth, delay, jitter and bit error rate.
- Optimize the throughput by introducing Performance Enhancing Proxies (PEP).
- Investigate suitable IPv6-based solutions for:
  - End-to-end Quality of Service (QoS);
  - Mobility;
  - Link aggregation techniques and the associated policy decision functions.

This paper will first give an overview of the network architecture that has been designed in section 2. Section 3 will describe the most important components in the architecture in more detail and will comment on the design decisions that have led to that architecture. In section 4, some of the performance results, obtained from a functioning prototype of the network architecture are discussed. Finally, section 5 formulates our conclusions and addresses future work.

2 Network Architecture Overview

Research on the topic of Train-to-Wayside-Communications-Systems has already been extensive [6], but most of the described architectures have resulted in high-level overviews so far [9]. In the TRACK project an architecture has been worked out in great detail, describing and implementing all elements.

2.1 TRACK Architecture

In figure 1, a general overview of the TRACK network architecture is shown. The left side of the picture represents the network architecture on board a train. In the TRACK architecture, this collection of components is referred to as "MCE" or Mobile Communications Equipment. Each train contains one MCE, as can be seen by the fact that multiple MCEs are stacked in the picture, representing multiple trains. In the picture, each MCE is depicted with two uplinks. This is not a requirement, however; each MCE can be equipped with as many or as few uplinks as is required. On the right side, the "WCE" or Wayside Communications Equipment is shown. Only one WCE is deployed in a control center. However, the WCE contains a complete set of components per MCE, as indicated by the stacked components in the WCE. As can be seen in the picture, the WCE only has one Internet uplink, as we assume it is connected via a broadband Internet

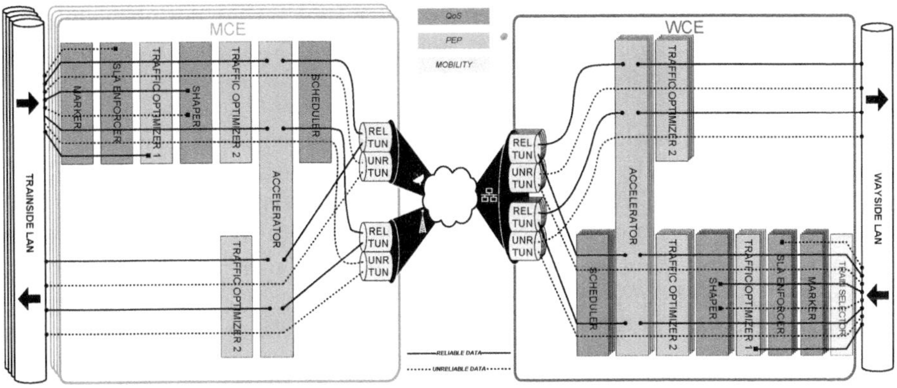

Fig. 1. TRACK network architecture overview

connection. The dimensioning of this Internet uplink is out of the scope of the work described in this paper.

Figure 1 uses a color coding scheme to indicate the function of each individual components: red components are responsible for QoS functions, green components offer a performance enhancing function and the blue components take care of mobility. Furthermore the TRACK architecture was designed with an end-to-end IPv6 based network in mind.

2.2 Implementation

Prototype Framework. One of the goals of the TRACK project was to have a functioning prototype version of the network framework by the end of the project. This prototype has been implemented using the Click Modular Router framework[8]. All of the components listed in figure 1 were implemented using one or more Click elements. Most of the elements were standard, while some of them have been custom made for the TRACK framework.

In order to design and implement the TRACK architecture, a number of existing mobility/addressing/multi-homing protocols have been investigated. For the performance enhancing elements, existing solutions have also been reviewed. Wherever possible, existing solutions or standards are preferred over custom solutions in order to maximize compatibility with existing systems.

Tunneling & Mobility. To support network mobility and multi-homing, which is required since every train has multiple uplinks that are not necessarily under the control of the train operator, some mobility solutions that have been taken into consideration are: Mobile IPv6[14], HIP[10], SHIM6[12], NEMO[5] and SCTP [19]. An important factor that was taken into account was the requirement that the end devices, whether they are passenger smartphones or diagnostic devices on board the train should not have to be modified in order to support mobility. Moreover, the mobility solution deployed needs to ensure that connections from an onboard device to the Internet are not affected by mobility and that the mobility solution should not break any of the optimizations introduced by the performance enhancing elements.

Performance Enhancements. The performance enhancing elements in the architecture can be divided in two classes: those elements that perform local optimizations (i.e. they can operate at the WCE or the MCE separately) and those that can only function if they are deployed both at the MCE and at the WCE. Performance enhancements can be made in numerous ways, such as deploying a local http proxy server (e.g. squid[18]) or a local DNS proxy; two examples of purely MCE-based proxies that do not require a WCE equivalent. Another possible performance enhancement is reducing the bandwidth required to transfer a given amount of data by installing a ZIP proxy[21] or employing robust header compression techniques like those described in [7], [4] and [3], which do need a counterpart at the WCE. A final optimization technique that has been considered, is the use of a TCP Performance Enhancing Proxy (PEP) such as SCPS[11].

2.3 Priority and Quality of Service

The architecture works with both the concept of priority and QoS. Every service type is categorized with a priority and a QoS class, thus it is possible that a service type has a high priority and a low QoS class or vice versa.

The priority level determines the importance of the service type: the higher the priority, the more important it is that data of this service type is sent through, the lower the priority the sooner packets from this service will be dropped when the available bandwidth is less than the needed bandwidth. Table 1 gives an overview of the eight different priority levels and which types of services are .

Table 1. Priority Definition

Priority	**Services**
1 (low)	Passenger Internet
2	Crew Intranet
3	Diagnostics
4	Application update Content Update TCMS Event
5	CCTV (security)
6	Intercom (VoIP) CCTV (safety) TCMS cyclic
7	Public address PIS data
8 (high)	Configuration traffic

Table 2. QoS Definition

Class	Delay	Jitter	Loss	Services
A	< 1 s	-	-	Passenger Internet Crew Intranet Diagnostics Application update Content update
B	< 0.5 s	-	-	TCMS event
C	< 1 s	-	$< 1 \cdot 10^{-3}$	CCTV (security)
D	< 0.07 s	< 0.016 s	$< 1 \cdot 10^{-2}$	Intercom (VoIP)
E	< 0.2 s	-	$< 1 \cdot 10^{-2}$	CCTV (safety)
F	< 1 s	-	$< 1 \cdot 10^{-6}$	TCMS cyclic
G	< 1 s	< 0.1 s	$< 1 \cdot 10^{-2}$	Public address PIS data Configuration data

On the other hand, the QoS class defines the requirements for a type of service to function properly. It does not take bandwidth into account, but does define bounds for delay, jitter and packet loss. We follow the IETF Differentiated Services (DiffServ) architecture[2] for traffic classification. Table 2 gives an overview of the different QoS classes that are used in our architecture. QoS and priority are not mapped one-on-one as such. However, if due to priority restrictions a flow with QoS class C-G encounters too much packet loss, that flow will be dropped because its QoS constraints can no longer be met. Similarly, even if there is enough bandwidth, but on links with delays and/or jitter higher than the delay or jitter allowed by the QoS class of the service, the flow cannot be set up.

3 Component Description and Design Decisions

As you can see in figure 1, the architecture is symmetrical for trainside and wayside, except for the fact that there is only one wayside network operating center for all trains. The components also fulfill the same functions on both sides of the architecture. In this section, the different components will be discussed and also the design decisions that were made in the project concerning these components. The components which handle mobility will be discussed in greater detail as they form the main topic of the paper.

3.1 Quality of Service Components

The Marker. The Marker is the first element in the chain for outgoing packets and is used to determine

- The flow ID of the packet
- QoS class of that flow
- Priority of that flow

A triple source IP address, destination IP address and Flow label in the IPv6 header define a flow uniquely and in our architecture the marker sets the flow label if the source did not do so. More information about the assignment of the flow label can be found in [13] and a description of the different QoS components and there implementation is worked out in [16] in more detail than in what follows.

The SLA Enforcer. The SLA (Service License Agreement) Enforcer will shape traffic flows according to the applicable SLA and drop or reduce maximum throughput for those flows for which the SLA cannot be met (e.g. when data volume limit has been reached). Thereafter the SLA Enforcer shapes all the flows of the same SLA together. The SLA is specified by VLAN tagging.

Shaper. The Shaper shapes the flows when there is more data to be sent than there is bandwidth available on the link. Different flows are handled appropriately considering flow priority and link occupation. The shaper receives the link occupation and the link-flow mapping from the Scheduler. If necessary, the Shaper drops packets with higher probability for low priority flows while trying to prevent starvation and to maintain a fair share policy for flows of equal priority.

Scheduler. The Scheduler decides which link to put the traffic on, based on matching service classes of the flows with delay and jitter properties of the link. The Scheduler also informs the Shaper of its decision, because the Shaper needs to know which flows belong to a link when it becomes overloaded. The Scheduler is based on Active Queue Management (AQM) [1], which means that there are two thresholds used for the queue: below the first threshold, the Shaper is permitted to allow more traffic going out. Between the first and second threshold, the Scheduler signals the Shaper to start dropping packets gradually. Starting on the second threshold, it is even possible that the Scheduler alerts the Marker to drop flows entirely to make sure that the most important flows are still coming through, meeting there QoS constraints.

The decision for the flow-link mapping is made based on the first packet and all consecutive packets of that flow are put on the same link to avoid jitter, unless the link goes down. In that case, the flow is rescheduled on a link that is still available.

3.2 Performance Enhancing Proxy (PEP)

The trains are connected to the wayside by means of costly media like 3G and satellite connections. Therefore it is important to optimize bandwidth utilization, which is taken care of by the PEP in the system architecture. In this section, the different components of the PEP will be discussed: two traffic optimizers and an accelerator.

Traffic Optimizer 1. As a first optimization along the way, the Traffic Optimizer 1 (TO1) tries to decrease the load on the wireless links by

- Caching information responses locally for requests it sees
- Responding with the cached information to future requests

Most of the functionality of the TO1 is situated on the application layer of the ISO OSI model [20] and includes caching proxies like web proxies, DNS caching, SMTP proxies. The caching does not only mean less load on the wireless network, but also a gain in response time for the user.

Traffic Optimizer 2. The second Traffic Optimizer (TO2) uses another technique to reduce the actual used bandwidth. The sender compresses the date, reducing traffic volume. The data is subsequently decompressed on the receiving end, so the eventual receiver perceives the data flow as unaltered. Compared to TO1, which tried to avoid sending traffic over the network by performing local caching, TO2 tries to minimize the amount of data sent over the network, reducing the bandwidth needed.

It is important to keep in mind that not all traffic should be taken under consideration for TO2: while e.g. HTTP and FTP traffic can be compressed well, data that is already compressed or appears to be random, like encrypted data do not get a benefit from TO2 and small packets like e.g. NTP protocol that are delay sensitive should not incur extra delay by compression.

Accelerator. Several long delay wireless connection types are used, especially satellite connections. It is well known that TCP does not react well to big bandwidth-delay products and therefore we have introduced the Accelerator in our architecture. It tries to solve the problem by

- Mitigating performance degradation resulting from the large round trip time between MCE and WCE
- Mitigating performance degradation resulting from multiple competing TCP connections on such links.

The Accelerator does this by breaking up one end-to-end TCP connection in three different connections: one TCP connection between the host on the train and the Accelerator there, then a non-TCP connection between the Accelerator on the train and the one at the wayside and thereafter a TCP connection again between the Accelerator on the wayside and the destination.

Both TCP connections are terminated at the Accelerators to give a perception of faster round trip time to the end hosts, allowing the TCP windowing mechanism to open the TCP window faster. Between the two Accelerators however, the data flow is converted to a "stateless TCP" stream, keeping the source and destination addresses and ports, but removing the congestion mechanisms of TCP, preventing competition between different data streams. The receiving end of the Accelerator sets up a new TCP connection to the destination of the packet and conceals the fact that the connection has been split by the Accelerators. In contrast to other TCP accelerators, this Accelerator is thus distributed between a sending and receiving module.

3.3 Network and Mobility

The last components have should be discussed, are the components that connect the train and the wayside to each other over the Internet, taking into account the (network) mobility on the moving train. To accomplish this, the TRACK architecture uses Reliable and Unreliable Tunnel elements, which will be discussed in this section. To set up these tunnels, the architecture uses SCTP (Stream Control Transmission Protocol)[19]. The advantages of using SCTP as a tunnel protocol can be described as follows:

SCTP can bundle reliable and unreliable streams into one SCTP stream with only one overall congestion control mechanism. This removes any contention between different streams. In a more standard network situation, it often happens that unreliable (UDP) traffic pushes the reliable (TCP) traffic away.

In many architectures, like e.g. VPN solutions it is sufficient to use an unreliable tunnel, since the reliable traffic inside the tunnel will take care of any packet loss, as it would do without the tunnel. It would even lead to performance degradation if reliable data traffic was put in a reliable tunnel. In the case of the TRACK architecture however, the Accelerator removes the reliability of the reliable connections between the accelerators at MCE and WCE. However, for both end points of the transmission, the connection should still be reliable or connections will break abruptly in the case of packet loss. Therefore, we need to re-introduce reliability. Since the SCTP protocol supports sending data reliably, packets that have been accelerated by the Accelerator will be transmitted in a reliable SCTP stream. SCTP will thus take care of any retransmissions and resequencing of packets. For unreliable traffic, there is no need to tunnel it reliably. SCTP offers the possibility to send data in an unreliable SCTP stream [15]; both reliable and unreliable streams are managed by a single SCTP connection, with a single congestion control mechanism.

In order to support mobility, the following scheme is used: one SCTP tunnel per link is set up by each train. Since the connection setup originates from the train to a fixed IP address on the wayside (the IP address of the WCE), no further mobility solutions are needed: the connections that are set up are the SCTP tunnel connections. Any data originating from a device on the train or on the Internet is unaware of the SCTP tunnel. This has the advantage that the end-to-end IP connections can function without deploying any additional mobility solutions. It is the responsibility of the SCTP tunnels to make sure data is exchanged between train and wayside. Every time a link comes up, the connection setup procedure is repeated.

Another advantage of SCTP is multihoming: a connection can survive link failures as long as at least one of the links is available. SCTP maintains a main path and several backup paths, with heartbeat messages going over each of them. If the main path fails, SCTP falls back on one of the backup paths. In the

TRACK architecture, link failure is also signaled to the Scheduler to ask whether the QoS constraints can be met. If due to link failure the available bandwidth gets too small, the Marker and the Shaper will act accordingly.

4 Results

4.1 Test Setup

As mentioned in section 2.2, a prototype framework has been implemented using the Click[8] framework. All building blocks shown in figure 1 were implemented using Click, with the exception of the "Traffic Optimizer" 1 and 2. In the prototype, these elements were left empty. As these elements provide well-known and mature functionality, such as proxies or header compression, they were not taken into account when testing the performance of the TRACK framework. If required, an existing implementation that provides these functions can be easily plugged in to the overall TRACK framework.

The main test focus was on the performance of Accelerator, Shaper, Scheduler and Tunnel modules. Most importantly, the cumulative effect of all these elements was tested to prove that they do not impact each other negatively when all of them are enabled. In order to test this prototype, an emulated setup was created using IBBT's Virtual Wall infrastructure. This infrastructure allowed us to emulate the network between a (virtual) moving train and a (virtual) wayside control center. All hosts ran Debian Linux, with SCTP support enabled. The TCP stack used is the standard Debian TCP stack, cubic TCP.

The test setup used is shown in figure 2; it consists of:

- Two *Train Hosts*, which are used to simulate onboard equipment;
- a *MCE* that has two Internet uplinks;
- *Impair*, the machine that is responsible for the network emulation. It adds and can constantly vary delay, jitter and, packet loss and varies the bandwidth between MCE and WCE;

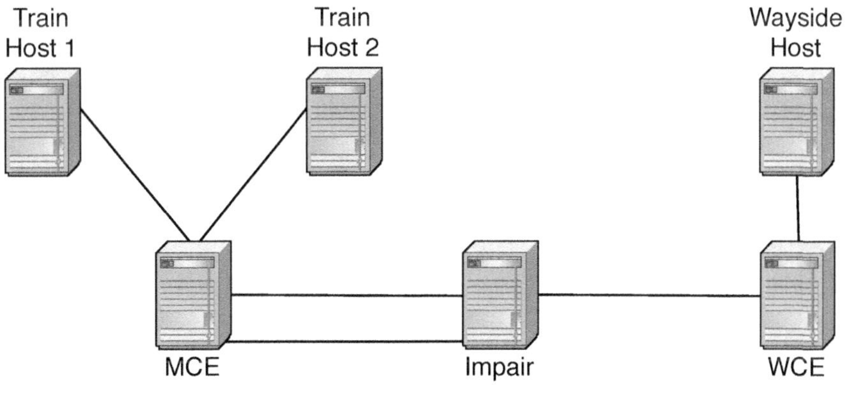

Fig. 2. Virtual Wall Setup

- a *WCE* that has only one Internet uplink;
- a *Wayside Host*, simulating equipment in the control center.

To test the performance of all components, various test scenarios have been defined. The most important ones are:

- Prioritizing some flows over others;
- Switching traffic flows from one link to another:
  - Switching between links with equal delay characteristics;
  - Switching from a low delay (10 ms) to a high delay (200 ms) link;
  - Switching from a high delay (200 ms) to a low delay (10 ms) link;
- Bandwidth and delay variation on a link;
- TCP accelerator effect on multiple TCP flows;
- SCTP tunnel performance.

4.2 Test Results

Flow Priorities. As an example of prioritizing some flows over others, figure 3 shows two (excerpts of) graphs, each containing two traffic flows. The blue line corresponds to a simulated CCTV video feed, which has high priority. The red line is a background bulk file transfer with low priority. The black line shows the total bandwidth used.

It should be noted that, when running this test, the link emulation was set in such a way that the link could not support both flows simultaneously at all times. In figure 3(a), no priority scheme was enabled. In figure 3(b), the CCTV video was prioritized over the file transfer. As can be seen in the results, the total amount of bandwidth consumed is roughly the same in both cases. However, in the second case, the CCTV feed consumes considerably more bandwidth, while the file transfer is using less. The variations in bandwidth used by the CCTV, as seen in figure 3(b), are caused by the Variable Bit Rate (VBR) codec used.

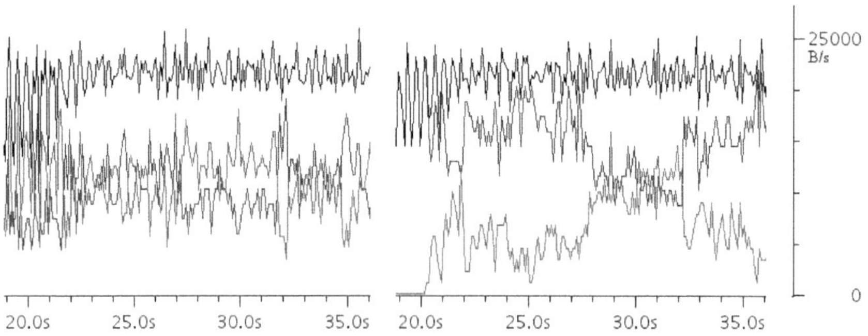

(a) Flows have the same priority (b) Flows have different priority

Fig. 3. Prioritization Test

The playback of both video feeds, which were recorded at the receiver, also show a very noticeable difference in both cases. While the first case renders the CCTV useless due to too much packet loss, the priority scheme ensures the CCTV video feed has sufficient quality to watch.

TCP Accelerator & SCTP Tunnel. In figure 4, two graphs are shown that demonstrate the effect of both TCP accelerator (PEP) and SCTP tunnel. The graph shows the throughput, in bytes per second, of three concurrent secure copy sessions (SSH over TCP) from a host on the wayside to a host on the train. The three sessions copy a 30 MB file over a 10 Mbps link that has a RTT (round trip time) of 400 ms. Each session was started 10 seconds after the previous one. Figure 4(a) shows the baseline scenario: both PEP and the SCTP tunnel have been disabled. Figure 4(b) shows the same test, but this time PEP and SCTP tunnel are both enabled. In both figures the black line represents the total amount of data on the link, while the red, blue and purple lines plot the individual SCP sessions. The data to generate these plots was gathered on the receiving host.

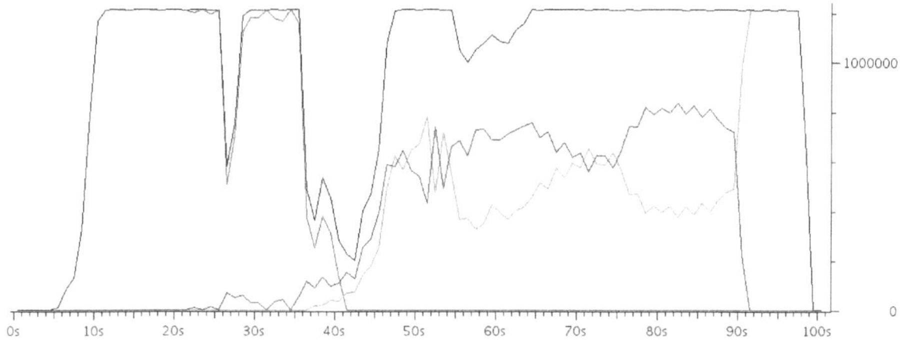

(a) TCP Accelerator and SCTP tunnel disabled

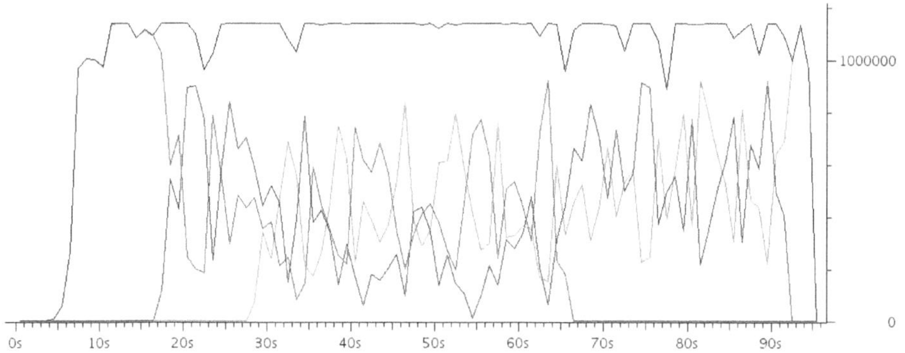

(b) TCP Accelerator and SCTP tunnel enabled

Fig. 4. TCP Acceleration and SCTP Tunneling Test

As can be seen in figure 4(a), in the baseline scenario, the second (blue) and third (purple) SCP sessions start very slowly. They only reach a reasonable throughput after the first SCP session stops. After that, it still takes some seconds before the second and third SCP sessions are able to fill the link, due to the TCP windowing mechanism operation.

When we enable both PEP and the SCTP tunnel, the result is shown in figure 4(b). Several effects can be observed:

1. The second and third session sessions start much more quickly;
2. Bandwidth utilization remains close to maximum, without large dips as in 4(a);
3. Each SCP sessions receives, on average, the same amount of bandwidth at all times. There is no longer one session that receives considerably more bandwidth than the others.
4. The total amount of time needed to perform the three SCP file transfers, is 5% shorter.

5 Conclusion and Future Work

In this paper, we presented an overview of the TRACK network architecture. This architecture has been designed to provide a reliable train-to-wayside communication architecture that can use multiple uplink technologies simultaneously and supports mobility without requiring modifications to the end hosts. To test the architecture, a prototype has been built using the Click Modular Router platform. This prototype has enabled us to run a multitude of simulations to test the functionality and performance of each individual component, as well as the architecture as a whole. The results of two of these experiments have been included in this paper as examples.

A follow-up project for TRACK called RAILS (Railway Applications Integration and Long-term networkS) has started on January 1st, 2012. This follow-up will build upon the TRACK network architecture, extending reliability and load balancing features, as well as integrating 3G/4G femtocells in the train coaches in order to increase in-train cellular network coverage. This introduces some challenges that need to be examined: in normal circumstances, femtocells are assumed to be placed stationary and have a fixed, high-quality Internet uplink. Onboard a train coach, however, neither of these assumptions still holds. While the train is moving, the radio environment around the femtocells continuously and rapidly changes. Moreover, the Internet uplink for the femtocells is also continuously changing in bandwidth, delay and jitter. These effects will most probably impact the performance of the femtocells. During the project, not only will the TRACK network architecture be extended, but full cumulative RF (Radio Frequency) exposure inside a train coach will be examined, most notably to compare the situation with/without cellular repeaters and with/without onboard femtocells. A final important network-related subject that will be treated in RAILS is the development of specialized handover mechanisms that take the specific characteristics of a train environment into account (e.g. the fact that a train follows a fixed route). The mechanisms developed could try to minimize the number of handovers, more intelligently choose a handover moment and introduce a self-learning process that dynamically builds up a database of preferred cells along a train route.

Acknowledgements. This research is partly funded by the Flemish Interdisciplinary Institute for BroadBand Technology (IBBT) through the TRACK project. The authors would like to thank all academic and industrial partners of this project.

References

1. Babiarz, J., Chan, K., Networks, N., et al.: Configuration guidelines for diffserv service classes. RFC 4594, Internet Engineering Task Force (August 2006), http://www.ietf.org/rfc/rfc4594.txt
2. Blake, S., Black, D., Carlson, M., et al.: An architecture for differentiated services. RFC 2475, Internet Engineering Task Force (December 1998), http://www.ietf.org/rfc/rfc2475.txt
3. Bormann, C., Burmeister, C., Degermark, M., et al.: RObust Header Compression (ROHC): Framework and four profiles: RTP, UDP, ESP, and uncompressed. RFC 3095, Internet Engineering Task Force (July 2001), http://tools.ietf.org/html/rfc3095
4. Casner, S., Jacobson, V.: Compressing IP/UDP/RTP headers for low-speed serial links. RFC 2508, Internet Engineering Task Force (January 1999), http://tools.ietf.org/html/rfc2508
5. Devarapalli, V., Wakikawa, R., Petrescu, A., Thubert, P.: Network mobility (nemo) basic support protocol. RFC 3963, Internet Engineering Task Force (January 2005), http://tools.ietf.org/html/rfc3963
6. Fokum, D., Frost, V.: A survey on methods for broadband internet access on trains. IEEE Communications Surveys & Tutorials 2, 171–185 (2010)
7. Jacobson, V.: Compressing TCP/IP headers for low-speed serial links. RFC 1144, Internet Engineering Task Force (February 1990), http://tools.ietf.org/html/rfc1144
8. Kohler, E.: The Click Modular Router. Ph.D. thesis, Massachusetts Institute of Technology (November 2000), http://read.cs.ucla.edu/click/click
9. Liang, X., Ong, F., Chan, P., Sheriff, R., Conforto, P.: Mobile internet access for high-speed trains via heterogeneous networks. In: Proceedings on IEEE PIMRC 14th International Symposium, vol. 1, pp. 177–181 (2003)
10. Moskowitz, R., Ricander, P.: Host identity protocol (HIP) architecture. RFC 4423, Internet Engineering Task Force (May 2006), http://tools.ietf.org/html/rfc4423
11. NASA: Space communications protocol standards, http://www.scps.org/
12. Nordmark, E., Bagnulo, M.: Shim6: Level 3 multihoming shim protocol for IPv6. RFC 5533, Internet Engineering Task Force (June 2009), http://tools.ietf.org/html/rfc5533
13. Pareit, D., Van de Velde, E., Naudts, D., Bergs, J., et al.: A novel network architecture for train-to-wayside communication with quality of service over heterogeneous wireless networks. EURASIP Journal on Wireless and Networking (2012)

14. Perkins, C., Johnson, D., Arkko, J.: Mobility support in IPv6. RFC 6275, Internet Engineering Task Force (July 2011), http://tools.ietf.org/html/rfc6275
15. Stewart, R., Ramalho, M., C.S.I., et al.: Stream control transmission protocol (sctp) partial reliability extension. RFC 3758, Internet Engineering Task Force (May 2004), http://www.ietf.org/rfc/rfc3758.txt
16. Rovcanin, M., Naudts, D., Pareit, D., Van de Velde, E., Bergs, J., Moerman, I., Blondia, C.: Data traffic differentiation and qos on the train, in fast parameter varying, heterogeneous wireless networks. In: Proceedings on 18th IEEE Symposium on Communications and Vehicular Technology in the Benelux (SCVT), pp. 1–6, 22–23 (2011)
17. Scalise, S.: Summary of conclusions and outlook esa artes 1 project internet to trains initiative, presentation. Tech. rep., ESA-ESTEC, Noordwijk NL (2008)
18. Squid: Optimising web delivery, http://www.squid-cache.org/
19. Stewart, R., Xie, Q., Motorola, et al.: Stream control transmission protocol. RFC 2960, Internet Engineering Task Force (October 2000), http://www.ietf.org/rfc/rfc2960.txt
20. Zimmermann, H.: Osi reference model-the iso model of architecture for open systems interconnection. IEEE Transactions on Communications 28(4), 425–432 (1980)
21. Ziproxy: the HTTP traffic compressor, http://ziproxy.sourceforge.net/

Conceptual Architectural Design of Indian Railway Intelligent Transportation Systems

Navin Kumar[1] and Nidhu Kumari[2]

[1] Institute of Telecommunication, University of Aveiro, Portugal
kumar@av.it.pt
[2] Tata Consultancy Services, Bangalore, India
nidhukumari@gmail.com

Abstract. Modification of railway transportation to railway intelligent transportation system (RITS) will not only compete from other modes of transportation but also increase the efficiency and safety. Speedy development of emerging technologies such as communication and computing, intelligent systems, technological development of railway industries can be effectively integrated over existing infrastructures for Indian Railway (IR) to develop and design RITS. This new generation of railway transportation (RITS) will able to meet the challenges of IR vision 2020 and beyond. The use of RITS architecture encourages structured development and integration of intelligent transportation systems (ITS) that lead to maximization of benefits by minimizing redundancies and maximizing capabilities. This paper discusses a conceptual distributed framework for a multi-layered RITS architecture that can be designed for integrating generated information. The paper main focus is conceptual design of RITS in IR network. It also addresses the design issues and challenges in the IR scenario.

1 Introduction

Indian railway is the third largest railway network in the world in terms of size, the world's topmost passenger carrier (in terms of passenger kilometers) and fourth largest rail freight carrier [1]. It is the backbone of India's infrastructure, along with the national highways and ports. It is estimated that more than 35% of the total freight traffic (tonne-kilometers) of the country moves by rail [2]. For certain core infrastructure sectors such as coal, power, steel, cement and other critical sectors like fertilizer, the share is even higher upto 70%. IR network binds the social, cultural and economical fabric of the country removing the distance barrier for its people. Its network of over 64,000 route-kms has integrated markets and connected communities over widely spread out geographies across the length and breadth of the country.

Despite all these, development in IR transportation system is very minimal. A mere 10,000 kms route has been added during last 60 years [1], maximum speed of passenger trains increased to 150 kmph, signaling and control systems remained vulnerable, accident rates remains high. IR is unable to meet the country's expanding travel and freight transportation needs. During major holidays and festivals, getting a

A. Vinel et al. (Eds.): Nets4Cars/Nets4Trains 2012, LNCS 7266, pp. 151–162, 2012.

railway ticket and making the trip are major events for travelers. There are some critical and important issues to be looked into: i) the requirement for capacity enhancement in the railway network, ii) technological up-gradation for better maintenance of railway assets, iii) need for enhancement to passenger services and safety, and iv) upgradation of the railway production units for improved efficiency and productivity. These shortfalls in IR have significant negative economic impact and need immediate attention from the authority. Some initiatives like project IR vision 2020 [3] show the government intention towards improving efficiency and safety of passengers.

Project IR vision 2020 call for a quantum jump in every dimension, breaking away from the path of 'incremental change' to one of the rapid growth. IR needs to restructure the management mode and upgrade the traditional technical approach. However, to realize development of railway transportation technology [4], there are many challenges:

- efficient and effective train management system, an intelligent train operation control system and intelligent inspection.
- railway disaster prevention, rescue and prompt response capability.
- supervision and control systems for the railway crossing and station including image identification technology.
- customers with real time information and inquiry service relating to the goods transportation
- to meet the demand of the inter-modal transportation.
- means for reduction in hazardous carbon emissions while promoting balanced sustainable development.

Railway transportation system therefore has to integrate organically large scale of the latest developments of science and technologies, making them as the brain and neural system of new generation of railway transportation system. Systems such as positive train control [5] in North America, Australia using the concept of communication-based train control, network centric system [6], advanced train management system with the aim of safety and increasing efficiency. Inclusion of technologies such as intelligent navigation system, intelligent emergency and safety supervision system, intelligent railway resource management system, intelligent train control and monitoring system, modern communication technology, data and image processing system, decision-making system, artificial intelligent networks, sensor networks, storage system and so on, the efficient and intelligent railway transportation systems can be designed. This will bring a revolutionary change in IR transportation system. Development of RITS is also useful for modern railroad system which will have integrated system of all transportation medium.

Following the terrorist events of September 11, 2001, and especially following the terrorist bombings of trains in Madrid, Spain, on March 11, 2004, in London, England, on July 7, 2005, and in Mumbai, India, on July 11, 2006, railway security has received increased attention. RITS will enable railways to prevent some types of terrorist incidents from occurring, and, should incidents occur, these technologies will enable railways to detect them, notify appropriate authorities, and recover more rapidly from the incidents. Railways will have continuous, real-time information with which they can manage their operations.

RITS can bring together a number of technologies, for example, sensors, communications, computing, intelligent control, management, interlocks, speed control, customer service, planning and scheduling and many more. This paper presents a conceptual design of a high level architecture for RITS in IR network. The paper highlights its importance and design issues. This conceptual design idea will carry many collective inputs to authorities and researchers in industries and academia. Together they can participate for innovative design of intelligent railway transportation systems especially in IR network.

Rest of the paper is organized as follows. A very brief review of accessible and existing IT infrastructure with IR is discussed in section 2. Major elements of RITS are presented in section 3, while its architecture is discussed in section 4. Important design issues and challenges are illustrated in section 5 and framework is addressed in section 6. Section 7 concludes the paper.

2 Accessible IT Infrastructure

Major parts of RITS would comprise IT related resources. This section therefore, identifies and evaluates existing associated infrastructure which will help in simple identification and integration of these resources to build system architecture for RITS in IR scenario. We broadly divide them into four segments: i) railway telecommunication, ii) railway computer networks, iii) passenger reservation systems, and iv) freight operation information system.

2.1 Railway Telecommunication (RailTel) Corporation of India Limited

Under the Ministry of Railways Government of India, RailTel [7] is responsible to create a nationwide broadband telecommunication and multimedia network. In fact, it has produced the state-of-the-art multimedia telecommunication network using SDH/DWDM (synchronous digital hierarchy/dense wavelength division multiplexing)-based transmission systems and high-end routers for MPLS-IP (multiprotocol level switching–Internet protocol) network. Existing 36,000 route-kms of Optical Fiber Cable (OFC) running along IR track in many parts of the country is planned to be extended to 50,000 km of rail-route covering most of the stations and commercial requirement (5000+) on its backbone by the end of 2012.

As shown in Fig.1(a), the network is designed as a layered structure for its effective management. Access Layer, Edge Layer and Backbone layer offer required capacity with desired granularity. Access layer provides (synchronous transport module level-1) STM-1 (155 Mbps) capacity at stations, spaced at every 8-10 kms. The traffic from access layer is aggregated on to the Edge layer having STM-16 connectivity, which is available at every 50-60 kms. Collecting the traffic from edge layer is finally aggregated on to the Backbone layer with DWDM/STM-16/64 available in big cities/towns situated at every 60-70 kms. MPLS IP core network connects 38 important points of presence (POP) locations all over India.

2.2 Railway Computer Networks (RAILNET)

The system structure of RAILNET is shown in Fig.1(b). RAILNET [8] offers computer connectivity between various offices such as Railway Board, Zonal Railways, Production Units, Centralized Training Institutes, and 46 Major Training Institutes. It has the potential for transfer of messages, files, e-mails between important locations on IR. In addition, the internal web site in railway board and zonal railway headquarters support codes, manual procedure orders, policy directives and so on.

2.3 Passenger Reservation System (PRS)

Center for Railway Information Systems (CRIS) is the internal service provider for IT. It is responsible for design, development and implementation of computer based information systems and associated communication networks and products, support and other systems. Currently over 1200 locations with 5000 terminals cover journey of more than 3000 trains and executes one million passenger transactions every day. It integrates five regional reservation centers as illustrated in Fig.1(c). Internet booking has started in 2002 and capable of providing reservations for 22 hours per day (uptime of PRS is 99.5%).

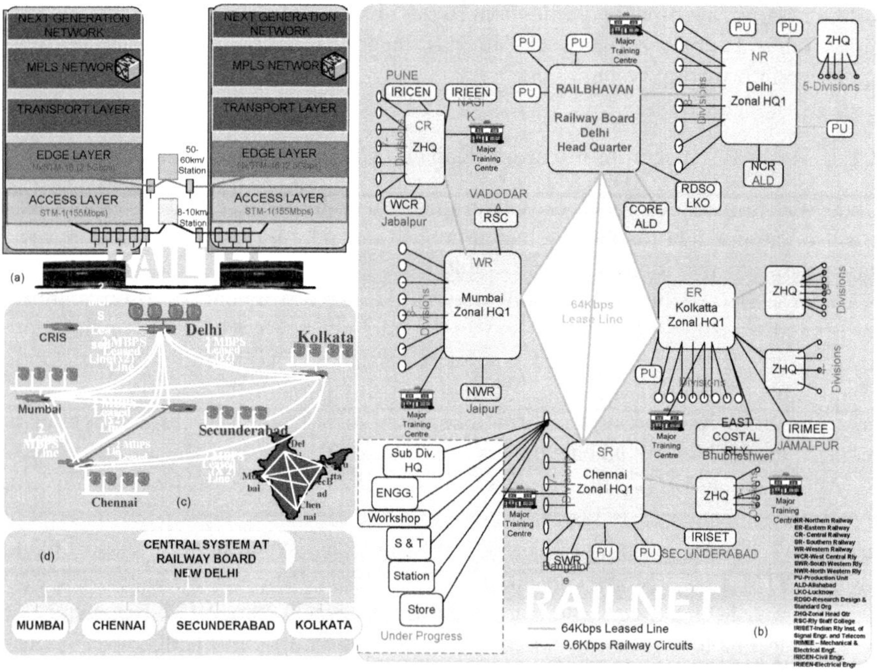

Fig. 1. Existing IT Infrastructures: a) OFC System Architecture (RailTel); b) Railway Network (RAILNET); c) Country Wide Network of Computerized Enhanced Reservation and Ticketing (CONCERT); d) FOIS

2.4 Freight Operations Information System (FOIS)

Responsible for total logistic system, FOIS deals with online real time system for management and control of freight traffic. FOIS is based on a hierarchical tree topology. An overall FOIS architecture is shown in Fig.1(d). It connects all major regional head quarter. It uses heterogeneous communication media such as IR owned OFC, digital microwave, leased channels from Bharat Sanchar Nigam Limited, Integrated Service Digital Network (ISDN), Very Small Aperture Terminal based satellite links. It constitutes Rake Management System which captures events such as customer indents, wagon supply, and departure/arrival in real time. Terminal Management System ensures billing and account of freight earnings, calculation of freight charges, generation of railway receipt and so on. Parcel Management System covers parcel booking, cancellation and loading, monitoring and tracking and Radio Frequency IDentification tags for tracing parcel. To meet the freight demand and enhance the quality of service, six dedicated freight corridors (DFC) are planned. Two of which are expected to be in operation before 2020.

IR is aiming for bringing technological changes in order to improve the system. The available resources with additional state-of-art IT infrastructure operational systems also provide a platform for RITS development. However, if broad design methodology and planning is considered at this level; efficient, effective and flexible RITS architecture can be designed and developed. Though, the IR vision does not include RITS development at present but it can be well introduced into the system with more flexibility to enhance the features in beyond IR vision 2020. In order to bring the conceptual idea of designing RITS, it is advisable to identify the key components of the architecture and its short falls.

3 Important Components of RITS Architecture

It can be conclude from above that the computerization on IR has been mainly in the form of data processing applications. However, a large complex infrastructure system of IR can benefit greatly from the intelligent use of IT. Available resources together with some additional intelligent technologies could make an efficient intelligent rail transportation system. The development of intelligent railway systems for command, control, communications, and information (C3I), as well as for braking systems, grade crossings, defect detection, and planning and scheduling systems can prevent collisions and over speed accidents. They can prevent hijackings and runaways, increase capacity and asset utilization, increase reliability, improve service to customers, improve energy efficiency and emissions. They can increase economic viability and profits, and enable railways to measure and control costs and to "manage the unexpected. Intelligent railway architecture will enable railways to improve their responsiveness to military deployments and to respond with flexibility and agility to rapid changes in the transportation marketplace.

3.1 The Major Constituents of RITS

There are four major domains in railway system: i) Rolling Stock Management, ii) Operations, iii) Traffic Management, and iv) Infrastructure Management. Apart from their dedicated operations they cooperate and interact to achieve the targeted end result which is to provide quality transport service for passengers and goods. RITS aims to be an enabling concept and architecture for improving the performance of each of the railway sub-systems, and the overall objective is to improve the system as a whole. RITS architecture is sure to bring in improvement by enhancing information exchange between the main processes through the players of the different subsystems or domains by means of more powerful, standardized and pervasive information and communication technology systems. We identify the major architecture components which operate at different levels of railway management and operation. This identification will allow us to understand their behaviors, their interactions, and usefulness and then it would be easy to integrate them into the whole system. The following components are grouped in:

- Intelligent Customer Service System (ICSS) – It is mostly customers/passengers/users oriented. It offers optimal travel choices. It helps freight shipment options. The system should include maximum of information such as ticket pricing, freight charges, booking information, train timings, real time tracking and so on.IR will be radically different from what it is today. ICSS can eliminate queues for tickets and congestion in coaches. Facilities like conferencing, banking and other IT-enabled services are not used in IR currently, which will be introduced in important trains shortly.
- Intelligent Train Control and Operation Management System (ITCOMS) – It deals with proper operation and control of train. It offers control over speeding up train and numbers of train on track with scheduling. It is also responsible for managing locomotive, rolling stocks and all operation related resources/infrastructures.
- Intelligent Monitoring and Emergency Rescue System (IMERS) – The objective of IMERS is to improve the safety of railway as well as passengers and goods. Various technologies and systems monitor real time status of trains, track surrounding environment and automate processing where applicable as well as keep informing track side supervisory and management systems. The track surrounding environment is not yet monitored. This leads to one of the major accident reasons as tracks are disturbed intentionally or because of other reasons. Passing through very difficult terrain and forest, track and its surrounding monitoring has become one of the key challenges to IR. Still there is no plan of monitoring the area near to the tracks.
- Reliable and Control Software – Control software should act as a nerve system for rail transportation. The software must be highly reliable and flexible to deal with the complexity of railway intelligent transportation system.
- Train-to-Ground Station Communication – This is another very important aspect to have effective control and operation of railway transportation system. It requires many heterogeneous networks for timely information transmission enabling immediate corrective and necessary action.
- Computational Platform – A computational platform can assess the reliability and safety of railway systems. It must be designed keeping in view the human-machine behaviours, management strategies and so on.
- Knowledge Discovery from Information Integration – Intensive information are available and must be collected, stored and studied on regular basis to bring improvement in rail transportation system.

These components operate at different levels in the layered RITS architecture. They must be interconnected with efficient communication (always on) facilities.

4 RITS Layered Architecture

Communication facilities are the key elements of RITS which keep every system integrated and coupled. As illustrated in Fig. 2, the layer architecture is divided into five different levels. The top level focuses on the major services (user/application subsystem) offered by RITS. While, rest of the main four levels comprises of all the key components, i.e. IMERS, ITCOMS and ICSS. It presents a functional view of the RITS user services [9] called as Logical architecture.

1. User Application Subsystem- This level is intended for end user applications such as reservations, enquiry, goods and freight booking and tracking.
2. The central management and control subsystem- It is responsible for overall management and control of all the available static and floating resources. They are responsible for both passenger and freight traffic management and control, safety and timely rescue operation. The user navigation system offers passengers or consignee making optimal travel plans by using available information round the clock. Similarly, e-Business allows Internet ticket booking, catering and railway related services via e-Payment. The intelligent resource management system helps efficient management of financial, infrastructure and maintenance. On the other hand, train operation and management system ensures train time schedule, passenger traffic management, freight traffic management and so on.
   Information exchange among different modes of transportation and decision about change in transportation mode is supported by inter model transportation system. The train monitoring and control system should able to detect position/status of train and other resources in real time. Because IR involves a large amount of transaction per day and a large number of employees, an efficient, comprehensive accounting and transaction system as well as reliable budget compilation system must be used.
3. Railway Station Subsystem– It consprises railway station inspecting system, signal testing and control system, interlocking control, freight transportation and user accessible services and applications. While inspecting system ensures proper order and present health of incoming and outgoing trains, signal testing and control takes care of error free and efficient routing and flow. Latest information relating to passenger and goods trains must be made available at the station. To avoid overcrowding automatic ticket machine and verification machine must be made available for passengers.
4. Train Onboard Subsystem– Very important in case of emergency. It should be equipped with inspection and maintenance systems for emergency. It should have effective train control mechanism to handle proper operation and communication system to deal with information transmission to railway station subsystem, or central management and control subsystem or train-to-train or other transportation

systems. A health management and information system becomes necessary in a huge geographical area.

5. Rail Trackside Subsystem–Keeping trains on track is an important issue. Intelligent track and road intersection supervision system will not only monitor the status and condition of tracks and associated infrastructure, provide the signal for incoming train but they will also automatically close the unmanned intersection points and prevent any disastrous that may occur. Wayside equipment sensors can identify defects that can occur on passing trains so they can be stopped for necessary repairs. Among the defects that can be detected by wayside sensors are overheated bearings and wheels, deteriorating bearings, cracked wheels, derailed wheels, and excessively high and wide loads.The subsystem also equipped with GIS based track maintenance, accident inspection and timely prevention and recovery mechanism.

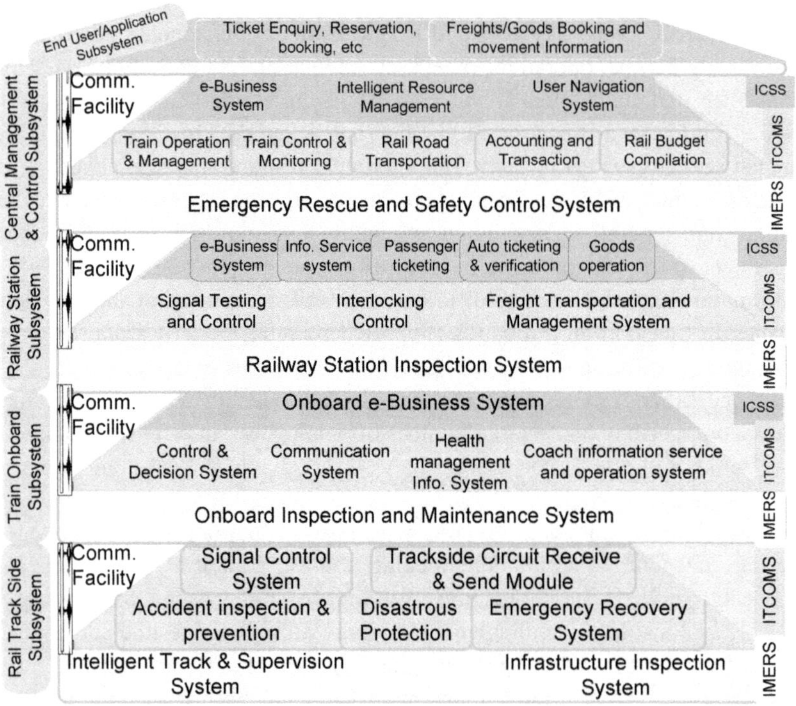

Fig. 2. Railway Intelligent Transportation Systems Architecture

5 Design Issues and Challenges

Railway system is very complex and produces large quantities of data, mostly in proprietary formats, which are difficult to understand, elaborate and share. The data is to be stored in efficient database systems from which useful information could be extracted. Efficient acquisition of information, transmission and flow and sharing of information are some of the critical issues which need to be addressed. An important method to integrate spatial and attributed data is to use common platform. Fig.3 illustrates an example of architecture for a common platform.

In a modular form the common data platform is divided into 3 level systems: (i) Ministry level – has highest access and full control management and planning, (ii) The Zonal Level –has full access and control of data in its respective zone, and (iii) The Divisional level – limited to the divisional geography. Data base design normally follows the Client-Server approach as in the case of largely distributed data in the enterprise like Wipro, Tata Consultancy Services (TCS), IBM etc. Requests from any end first go to the "listener", few thousands of listeners handle the multiple requests in the beginning and they transfer the request to server, and in the server there are three phases: i) Parsing, ii) Fetch, and iii) Execute. Parsing will check the syntax of the requests. Fetch is a normal process to return the answer of the request and execute option will find the least approach to return the result of the request. Least approach can get the result in minimum time of interval following different approaches. Within few main data bases, there will be logical division of main data base like train status, ticketing, parcel information, rescue and recovery and so on. The provision of back-up is available for defined time duration of current transaction. There can be 3-4 mirror copies of back- up data base at different locations for recovery in the case of crash and such events. Thus a huge and efficient data base system is mandatory requirement. This will establish the standard of structures and transmitting methods for railway transport and build a series of data warehouse used for organization, storage, query and communication of common information. Designing this data base is a big challenge. Apart from this, there are many important design issues that need to be addressed in order to develop the architecture. Some of the basic issues are:

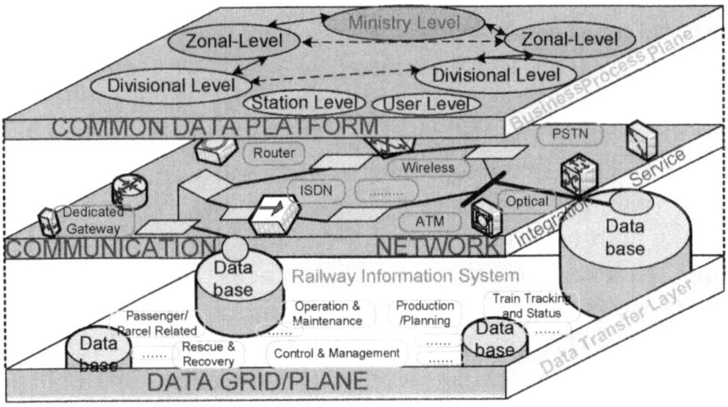

Fig. 3. Architecture for a Common Platform

- Standard infrastructure – IR need to improve the very basic railway infrastructure such as electrification, quality tracks to be used by high speed trains, engines and coaches equipped with modern IT, communication and control devices.
- Distributed Sub-systems - Because many sub-systems are scattered and distributed in a large geographical region, monitoring and integrating distributed information

of infrastructure and rolling stock data sources in the railway domain, stakeholders, hardware and software systems are very difficult.
- Process Coordination - Coordination of distributed heterogeneous processes is multifarious. Similarly, the cooperative computing performance data from infrastructure, operational management, rolling stock and traffic management become complex.
- Network Integration - Integration of heterogeneous networks technologies is one of the most challenging tasks.

RITS plays very important part in obtaining this information. In addition, development of RITS will enable effective usage of existing systems. It will be flexible enough to be capable of updating at any stage and any time in the future.

6 Design Framework

A well organized and careful planning is required to design RITS architecture in IR network in order to enhance the performance such as safety, capacity, punctuality, speed and so on. Using intelligent monitoring, maintenance, decision support, key performance indicator (KPI), better data interpretation and evaluation, flexible data and process integration can enhance the performance of RITS. For this, system should require co-operation, heterogeneity and efficient distribution. An intelligent communication system with many different flexible communication networks can provide ubiquitous data access. Communication networks offer dependable connections to the wayside and central data sources. It can extend reliable service transport to mobile resources. This will result in optimization of railway system's observability and controllability. The system architecture defines the functions that must be performed to implement a given user service, the physical entities where these functions resides, the interfaces/information flows between the physical systems and the communication requirements for the information flows. It also defines the requirements for the standards needed to support national and regional interoperability.

The process of building RITS architecture starts with clear definition of user needs. Then the system specifications must be defined which follows the design. On systems integration, testing and validation are carried out. A methodology adopted in the project InteGRail [10], intelligent Integrated Railway in European context, can be used following combined approach of top-down and bottom-up. The top-down approach defines a high-level description of the railway system in terms of KPI. The bottom-up approach examines the current practices and requirements. Available technologies then can be examined to convert data into information and allow them process intelligently. The processed information is then stored into suitable repositories and processed to obtain necessary knowledge. The knowledge is made available by means of suitable services such as, following service oriented approach. The methodology is expected to result in development of system architecture, complete and flexible enough to allow the implementation of all envisaged applications, in the framework of their respective scenarios.

Complexity of the system design can be reduced by subdividing the major operations. For example, a devised framework consisting of four major phases/tasks of design is shown in Fig.4. Different divisions may operate simultaneously; however, some are dependent on the completion of others. Some continuous processes like overall management, evaluation and co-ordination between scientific and industry community, providing support, cost analysis and training go on throughout project life cycle. It includes the platforms to provide sufficient dissemination and preparation of training courses, as well as the provision and the application of a harmonized methodology to calculate the economic impacts.

The system architecture defines the functions that must be performed to implement a given user service, the physical entities where these functions resides, the interfaces/information flows between the physical systems and the communication requirements for the information flows. It also defines the requirements for the standards needed to support national and regional interoperability.

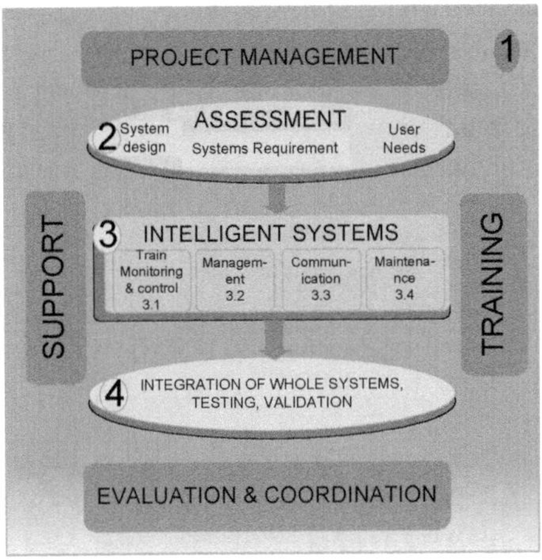

Fig. 4. Architecture Design Framework for RITS

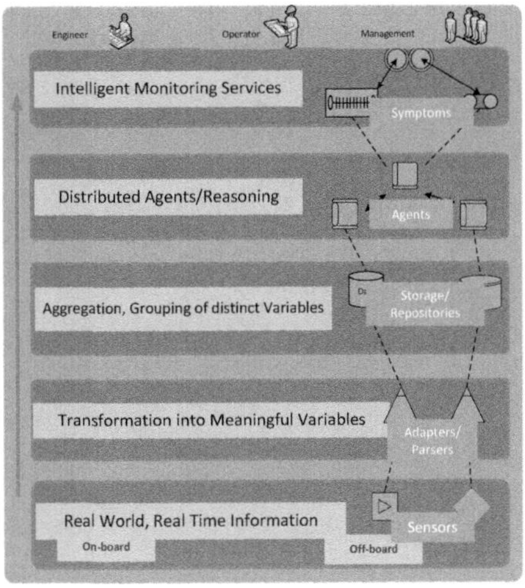

Fig. 5. Advanced Monitoring Systems for Intelligent Railways

RITS, a new generation of rail transportation should incorporate advanced and intelligent subsystems such as advanced intelligent monitoring and control systems for train, infrastructure, supervision; advanced maintenance and so on. An example of standard architecture for a new generation of monitoring and control systems is shown in Fig.5. It is expected that all needed devices can dynamically plug-in, with simple decentralized intelligent re-configuration in case of service update, changes or other modifications. It should have: the ability to expand and dynamically accommodate to new functionality (e.g. new sensors) via service discovery; to be able to upgrade, and to guarantee that changes will be consistent in functionality for other autonomous components. Similarly, intelligent maintenance ensures availability, reliability and safety. It should include methodology to optimize and integrate process to improve the efficiency of maintenance operations and reduce waste (of material, capital and resources). For example, lean maintenance optimizes and improves efficiency, linking of maintenance planning into the railway operations process balances critical needs of two areas minimizing wastes. Automation of manual activities and integration of support systems reduce the time needed for maintenance and prevent mistakes, thereby improving vehicle availability and reliability.

7 Conclusion

Design and development of intelligent railway transportation systems becomes extremely important especially in scattered geography and huge networks such as in Indian scenario. However, to achieve this mile stone a proper planning and study must be undertaken. Though, there are many research and training organizations supporting Indian railways helping in engineering, design and building IR as one of the competitive transportation systems, but RITS merits can't be ignored. RITS integrate all the components in a most efficient and optimized manner. Therefore, the proposed RITS architecture will able to meet many of the challenges set by IR vision 2020 and beyond, making it one of the most developed and efficient transportation systems in the world. Many challenging issues have been discussed. A framework of design is also presented. It is expected that, such conceptual knowledge will not only facilitate Indian railway's research and training organization but also accelerate the railway development in rest of the world.

References

1. Planning Commission, Govt. of India, Total Transport System Study, RITES (A Govt. of India Enterprise) (2009)
2. Indian Railways, http://www.indianrailways.gov.in/indianrailways/White%20Paper_Eng_SUBMITTED%20To%20Parliament.pdf
3. Indian Railways, http://www.indianrailways.gov.in/indianrailways/VISION%202020_Eng_SUBMITTED%20TO%20PARLIAMENT.pdf
4. Jia, L.-M., Li, P.: The sysetm architecture of Chinese RITS. Proceedings of the Eastern Asia Society for Transportation Studies 5, 1424–1432 (2005)
5. Petit, W.A.: Interoperable Positive Train Control: Existing Safety-Critical Architecture. IEEE Vehicular Technology Magazine (December 2009)
6. Ditmeyer, S.R.: Network Centric Railway Operation Utilizing Intelligent Railway Systems. Journal of Transportation Law, Logistics and Policy, Third Quarter 77(3) (2010)
7. http://www.railtelindia.com/Infrastructure.aspx
8. Kumar, A., Gadgil, A., Satapathy, A., Upadhyayula, R., Prabhudesai, S.: Indian Railway IT Interface, Report on Infrastructure Development in Financing, IIM Ahmedabad (2001)
9. Li, P., Jia, L.-M., Nie, A.-X.: Study on Railway Intelligent Transportation System Architecture. In: IEEE Proc. Intelligent Transportation Systems, vol. 2, pp. 1478–1481 (December 2003)
10. Intelligent Integration of Railway Systems, http://www.integrail.info/

Automatic Alert Generation from Train to the People at Unmanned Level Crossings Using Principles of IoT

Abhinav Jha, Amit Kumar Agrawal, and Chirabrata Bhaumik

Tata Consultancy Services
Plot A2M2&N2, Block EP, GP
Saltlake Electronics Complex, Sector V
Kolkata - 700091
{abhinav.jha,amitk.agrawal,c.bhaumik}@tcs.com

Abstract. This paper presents a novel concept of an application to improve safety at unmanned level crossings without the burden of a huge infrastructure cost to the railways governing body. The application will keep track of all running train locations and the unmanned level crossings on its way. The application needs to have access to location Appplication Programming Interface(API) provided by Telecom Service Providers. With usage of these APIs the application will be able to keep track of all the people near unmanned level crossings. All service providers have Bulk Short Message Service(SMS) APIs using which the application will be able to send alert messages to people in vicinity of unmanned level crossings. It will make road users aware of the train approaching the level crossing and reduce the risk exposure. The same principle of Internet of Things (IoT) can be extended here for developing a query service by users about approaching trains in a particular level crossing.

Keywords: Unmanned Level crossing; Connected Train, SMS, SOS, IoT, Warning System, Safety, Road users, Grade crossing, train to infrastructure communications; traveller information services.

1 Introduction

In developing countries like India, accidents at unmanned level crossings are still a big challenge where the train network span length and breadth of the nation having 32,735 level crossings out of which 14,896 are still unmanned. According to the data, 229 people were killed at unmanned level crossings in 2010-11 and 95 in 2011-12 due to rail accidents [1,9]. Even though, the technology has improved and there are sensor-based solutions for automatically controlling level crossing gates, these solutions are often difficult to implement because of huge infrastructure cost involvement. It is difficult to ensure the maintenance and protection of devices and sensors at secluded unmanned level crossings in these countries.

A. Vinel et al. (Eds.): Nets4Cars/Nets4Trains 2012, LNCS 7266, pp. 163–173, 2012.

The main motivation behind this application is to build a warning system, which does not require any hardware installation at level crossing locations, or on the railway track. Moreover, the alert raised by the application must be delivered reliably to at least some people near the level crossing.

To meet the objectives mentioned above, we have devised an application, which does not require any extra hardware installation at the level crossing locations or train unit. It blends existing technology components to build a solution, which meets the stated objectives. It is an IoT based application, which just needs a smart phone (with Global Positioning System (GPS) module & General Packet Radio Service (GPRS) connectivity) mounted in the train unit. The smart phone in the train unit acts as a geometrical location sensor and keeps on posting its geometrical position, speed, direction etc. to the application. We have specially mentioned smart phone here as it has become a commodity item available in every day market and its price will decrease to a large extent with the explosion of new smartphones every day with GPS module and GPRS modem. Alternatively the smart phone can easily be replaced by a specialized hardware with a very reduced BOM (Bill of Materials) cost with the necessary processing unit, sensors and a modem module. To manage the data sent by the train unit, the application uses Sensor Observation Service (SOS).

The application keeps the train routes, and level crossing locations as static data in a spatial database. On receiving a train unit whereabouts, the application projects the warning zone based on the trains speed and location, and identify all the level crossing locations lying within the warning zone.

The application needs access to mobile operator services in order to raise SMS alerts (considering co-operation, policy mandates & investment from the government and mobile operators) for the level crossings lying in the warning zone. The alert service will only be provided to subscribers in order to honor the privacy & consent of the user and avoid legal complications. As mobile phones are quite common these days and within the reach of everyone, there is a high probability that SMS alert will be able to caution majority of the people near the level crossing. The alert will be a subscription based service and people can SMS to a certain number to deactivate and activate this service at their discretion.

In this paper, we explore the previous work done in this area and its relevance to our concept in section 2. In Section 3, we describe the System architecture, components and their interaction. Section 4 provides the functional details of various components along with the algorithms in form of pseudo code. Section 5 shows the results and screen shots visualizing the concepts. Section 6 provides the concluding remarks.

2 Previous Work

Following are the major works, which are close to the subject area of this paper:

– [2,4] depends on hardware (transceiver) installed at trains/tracks/level crossing. It neither employs a server-based system nor gives SMS alerts. The decision for alert generation is taken at the end user vehicle.

- [3] Describes a system where a locomotive signals its approach to the upcoming railroad crossing signals which in turn activate warning devices. It does not communicate with a central server, and does not push SMS alerts.
- [5]Describes an approach, which replaces signals and sensors on the route by radio communication and software computations. It is flexible and cost effective approach and is closer to our objective of creating an infrastructure less solution. It drifts away from our objective of providing the SMS alerts, by communicating directly with the level crossing to raise an alarm.
- [6] Depends on ultrasonic-based obstruction detection at a level crossing with snow covered tracks and discusses the safety implications.
- [7] Targets the performance of warning systems required at level crossings for soundproof vehicles, high speed conditions and poor visibility due to weather conditions. It suggests directional sonic devices and tactile warning devices, which gets activated on a real-time closing velocity determination.
- [8] Covers an extra mile and discusses the second level of protection, i.e. additional measures required to mitigate the intrusion incidences by providing real-time crossing site image to the train drivers.

3 System Overview

Fig. 1 illustrates the basic architecture of our proposed application. It comprises of a train unit, a central application server and telecom operator services as major components, which are described in the following subsections.

3.1 Train Unit

The train unit needs a smart phone with a GPS module and GPRS connectivity. The smart phone identifies its speed, direction and geographical location (latitude, longitude) using the GPS module and sends the data at periodic intervals to the central application server through GPRS connection.

3.2 Central Application Server

This is the back end server meant to receive, store and analyze all the data posted by train units identify the warning zone and concerned level crossings, identify the registered mobile users in the warning zone and raise the warning alerts to them. It has following modules

SOS Module. SOS is a standard provided by Sensor Web Enablement (SWE) initiative of Open Geospatial Consortium (OGC). It supports spatial data type and provides XML interface to store & fetch sensor observations. The authors have used open source implementation of SOS from 52North.org and it was available as an offering for their IoT platform. It has a generic schema, which can accommodate the sensor observations from any type of sensor, and does not require a schema redesign. The proposed application treats every train unit as a sensor, and store the data posted by them as a sensor observation in a spatial database. The static train routes and level crossing locations are kept in the spatial database. On every insert, the database trigger checks the geographical location & speed of the train unit and calculates the warning zone. It notifies the traffic router module about the warning zone and the level crossings lying on it. The Results section contains screen shots illustrating the warning zone calculation.

Mobile Locator Module. It provides an interface for registering a mobile number to receive warning alerts, and store them in the database. It can locate the mobile phones in vicinity of a given geographical location by calling the API provided by telecom operator.

Traffic Router Module. It keeps on listening to the notifications from database trigger for the level crossing location lying on the warning zone. On receiving a notification, it requests a list of registered mobile numbers in vicinity of notified level crossing locations from mobile locator module, and push an SMS alert to them by calling bulk SMS API provided by telecom operator.

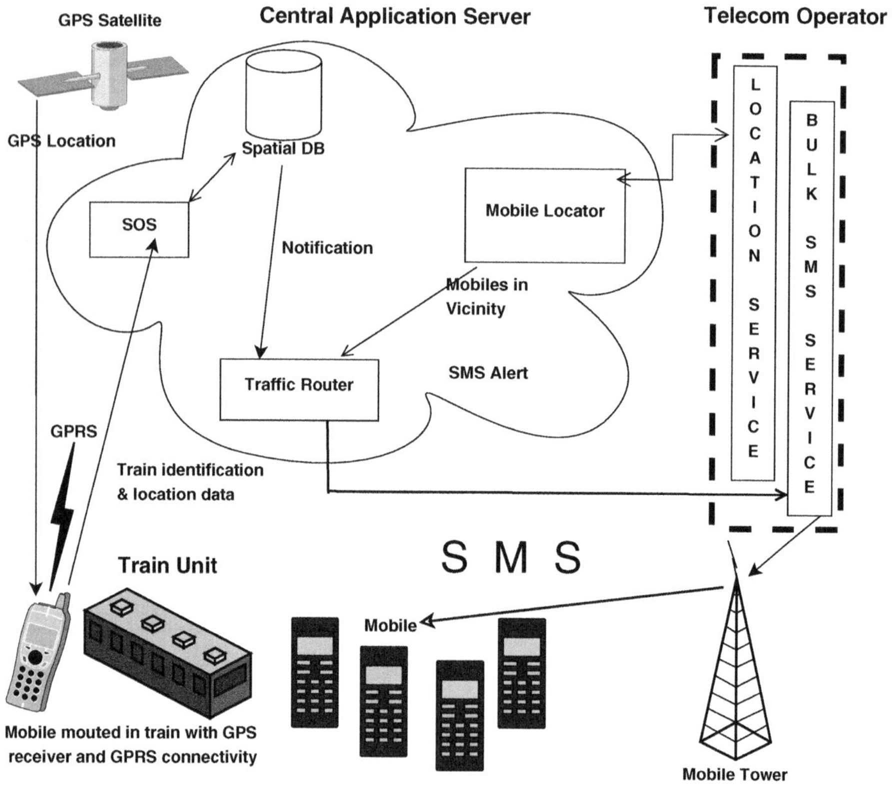

Fig. 1. System Architecture & Overview

3.3 Telecom Operator

This component lists the APIs provided by the telecom operator. Location Services API and Bulk SMS Services API are the major offerings of this component. Mobile locator module calls Location Services API to locate the mobile phones in vicinity of a given location. Bulk SMS Services API is used by the traffic router module to push warning alerts.

4 Functional overview

This section traces the flow of data through various components of the system and its culmination in the warning SMS.

As GPS & GSM modules are not integrated in a train unit, this component is prototyped as an iPhone application as shown in Fig.2. iPhone has been chosen as a train unit in our application for the trial run as performance wise we have found iPhone to be the best in terms of sensitivity of its sensors as well as reliability. Train number, upload frequency and the central application server location can be configured using settings screen. The application screen shows the current value for speed, geographical location and direction parameters for the latest post. The data originates at the train unit and flows to the SOS module of the Central Application Server as SOS compliant XML. The SOS module parses the XML data and inserts in the spatial database. A pre insert trigger checks the data and finds out the warning zone. A snapshot of a typical XML is as follows.

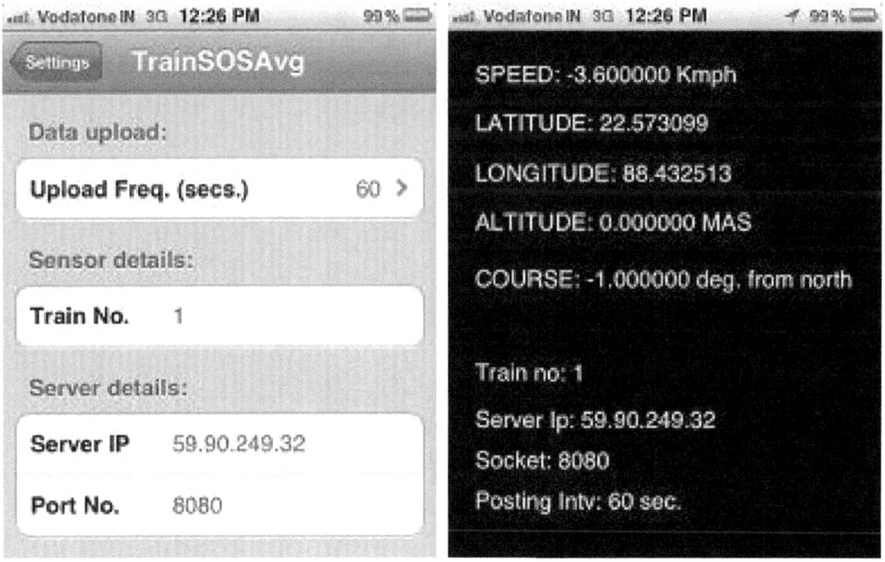

Fig. 2. Train Unit iPhone Application

```
<InsertObservation xmlns="http://www.opengis.net/sos/1.0"
  .....
  <swe:CompositePhenomenon gml:id="cpid0" dimension="1">
    <gml:name>resultComponents</gml:name>
    <swe:component xlink:href=
        "http://www.opengis.net/def/uom/ISO-8601/0/Gregorian" />
    <swe:component xlink:href=
        "urn:ogc:def:phenomenon:OGC:1.0.30:TrainNo" />
    <swe:component xlink:href=
        "urn:ogc:def:phenomenon:OGC:1.0.30:Speed" />
    <swe:component xlink:href=
        "urn:ogc:def:phenomenon:OGC:1.0.30:Degree" />
  </swe:CompositePhenomenon>
  .....
  <gml:Point>
    <gml:pos srsName="urn:ogc:def:crs:EPSG::4326">
        52.1 8.1
    </gml:pos>
  </gml:Point>
  .....
  <swe:values>
        2011-04-03T04:45:15+00,TrainFoi051,504,51,225
  </swe:values>
  .....
</InsertObservation>
```


The psuedocode of the warning zone calculation algorithm is as follows.

```
INPUTS :  TrainRouteId, TrainLocation,
          TrainSpeed(kmph), TrainDirection, Time
PRECONFIGURED : THRESHOLD_TIME  (120 sec)
Static Routes : Route_Id, Route,
                Level_Crossings, Route_Length
BEGIN

    /* Fetch static route */
    TrainRoute := Select Route
                Where TrainRouteId = Route_Id
    TrainLevelCrossings := Select Ulc
                Where TrainRouteId = Route_Id
    TrainRouteLength := Select Route_Length
                Where TrainRouteId = Route_Id

    /* Convert THRESHOLD_TIME to THRESHOLD_DISTANCE */
    /* using Speed */
    THRESHOLD_DISTANCE :=
          (TrainSpeed * 1000 / 3600) * THRESHOLD_TIME
```

```
    /* Calculate Warning zone as Linestring */
    /* using spatial queries written as stored proc */
    WarningZone := warningZone_stored_proc(TrainRoute,
                TrainLocation, THRESHOLD_DISTANCE)

    /* Identify LevelCrossings on warningZone */
    /* using spatial queries written as stored proc */
    LevelCrossings := LevelCrosing_stored_proc(
                WarningZone, TrainLevelCrossings)

    RETURN LevelCrossings
END
```

For each train number, corresponding route as line string and level crossing locations as multipoint are stored in a static table. A pre-configured threshold time value is used in combination with train speed to calculate the distance to be marked as warning zone. This enables the application to raise an alert within safe distance from train unit, removing the chances of road user getting distracted by the alert SMS. On receiving the list of level crossings via the algorithm depicted above, the trigger notifies the list of level crossings to the traffic router module.

The pseudo code representation of traffic router module is shown below. It depends upon mobile locator module to get the list of mobile numbers in vicinity of a given location or a list of locations.

```
INPUTS : LevelCrossings
BEGIN
  WarningMessage :=
          "Train is arriving shortly at this level crossing"

 /* Query mobile numbers from mobile locator */
 MobileNumbers :=  mobile_list_from_mobile_locator(LevelCrossings)

 /* Issue SMS using telecom operator Bulk SMS API */
 Push_SMS_using_Bulk_SMS_API(MobileNumbers, WarningMessage)
END
```

The mobile locator module is responsible to provide mobile registration facility to the target application users and stores them along with the geographical locations and timestamp. The Location Services API provided by telecom operator is based on mobile number. To simulate the location based query, it caches the mobile locations in one thread periodically and provides service to traffic router module in another thread.

The inner working of the overall scheme is summarized as activity chart in Fig. 3.

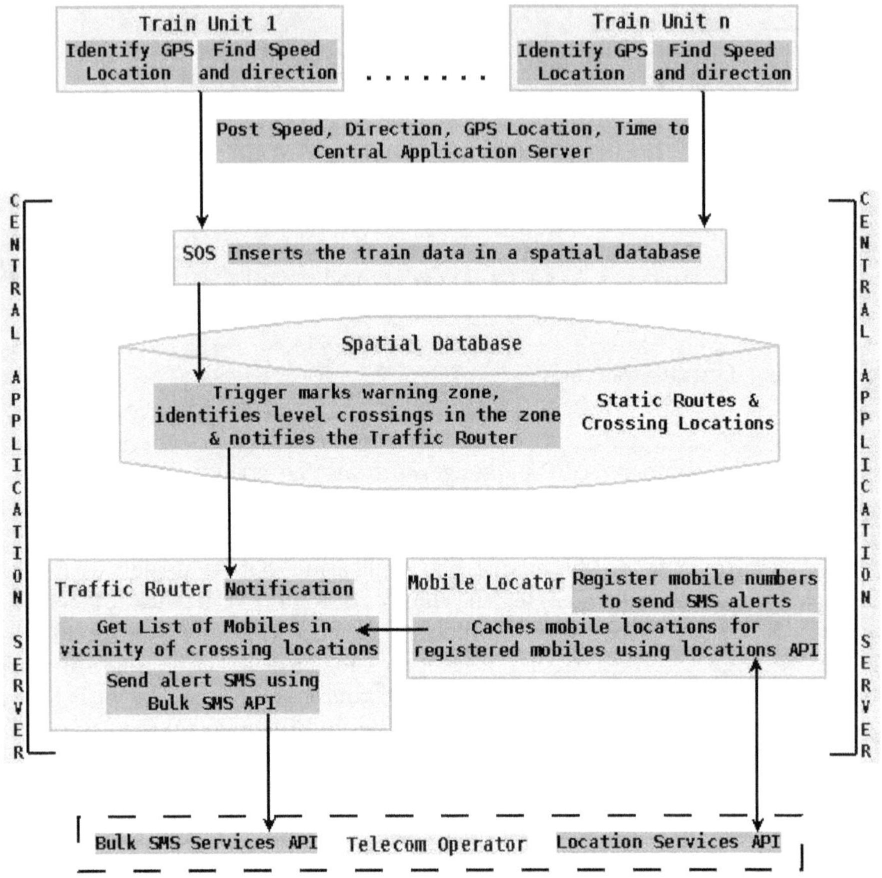

Fig. 3. Activity Chart for Overall Working Scheme

5 Results

In this section, the final output & intermediate results from our application is mentioned as screen shots taken from visualizations of routes and SMS. The first screen shot as shown in Fig. 4 refers to the warning SMS generated by using Bulk SMS API provided by the Telecom Operator. This is the final output from our application, meant to serve as a warning to the people in vicinity of the level crossing.

The train route and level crossing locations corresponding to each train number is stored as static table in the database. Since performing this test is difficult, a simulator has been developed using google maps with the roads acting as railway tracks. We have made some well-known road crossings as level crossings. The train has then been simulated to run on the google map with posting of its location to our backend using google location points. This has been done for testing the backend system. Fig. 5 represents one such static route marked as

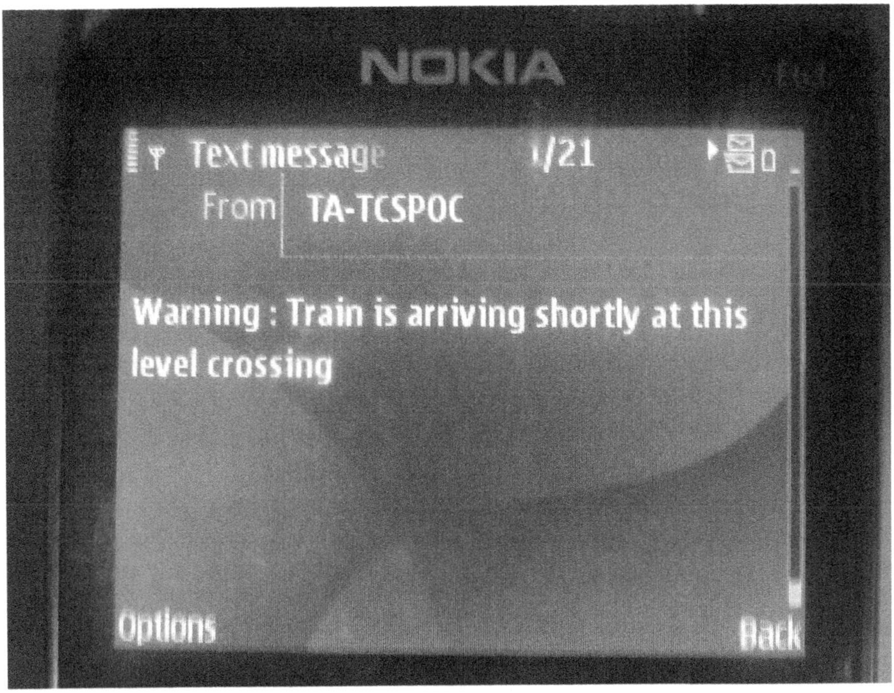

Fig. 4. Snapshot view of SMS

Fig. 5. Train Route and Level Crossings Marked on a Map

a green line on the google map. The train is positioned at the beginning of the route. The level crossings are also marked along the whole route. The view of Fig. 5 is a representation of a dashboard application, which can be hosted on a centralized train control room. This kind of view will be very helpful for railway authorities, as it gives them snapshot of all running trains under their authority.

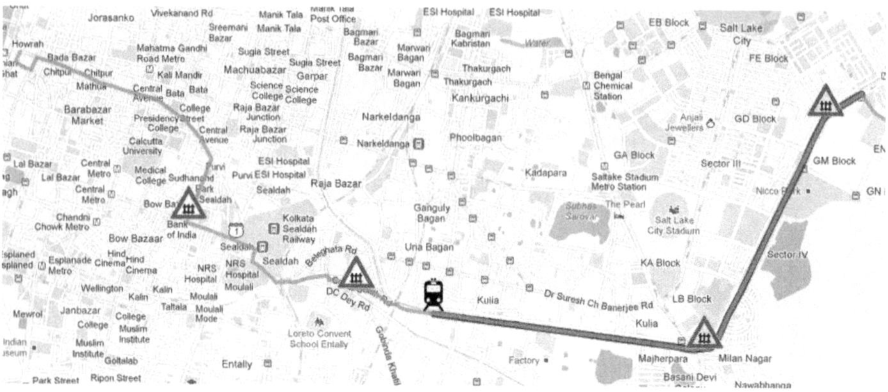

Fig. 6. Train Route and Level Crossings with Highlighted Warning Zone

While processing the data posted by train unit, current position of train is considered to calculate the warning zone, and all the level crossings lying on warning zone are also identified. Fig. 6 marks the train current position of the train. The highlighted part of the route represents the warning zone. The level crossings lying on warning zone are also highlighted. This is the intermediate result based on which warning alert is raised.

Field tests have been carried on the marked routes simulating train with a car, track with road and road crossings as level crossings demonstrating the practical implementation of this application.

5.1 System Latency

For the sample tests conducted, the SMS API response time is quite fast, and the SMS gets delivered within 5 to10 seconds. However the Location API provided by telecom operators is based on mobile numbers. The Mobile Locator module needs to cache the mobile locations every minute and simulate the location-based query. The database trigger processing is quite fast and does not introduce any considerable latency in the system.

6 Conclusion

Thus we have devised an infrastructure less system, which can generate SMS warning messages and alert the road users about the approaching train near level crossings. It thus acts as a first level of defense near secluded unmanned level crossing locations.

There are a few areas, where improvement can be made & system latency can be reduced. The Location Services API must take location as input and provide list of mobile numbers, so that the locations of mobile phones is more precise. The trigger mechanism may not be appropriate for large volume of traffic. Streams processing might be a viable alternative to handle it.

This concept can be extended to provide traveler information by introducing a query interface. The application users can query a train position nearest to any given level crossing or request the status of any level crossing by providing its geographical location.

Acknowledgements. The authors would like to sincerely thank Mr. Arpan Pal for reviewing the paper. The authors would also like to thank Mr. Debnarayan Kar, Mr. Kallol Ghosh, Mr. Prasun Bhattacharjee, Ms. Priyanka Sinha and Mr. Joydip Das, all members of CTO Innovation Lab Kolkata for helping in implementation of the pilot deployment of the project.

References

1. Balchand, K.: Safety at level crossings remains a problem, THE HINDU, New Delhi, June 22 (2010)
2. US 5554982: Patent titled Wireless train proximity alert system by Hughes Aircraft Co.
3. US 5620155: Patent titled Railway train signalling system for remotely operating warning devices at crossings and for receiving warning device operational information by Michalek; Jan, K
4. US 6241197: Patent titled Automated Railway Crossing by Harland; Sydney A
5. Reif, W., Schellhorn, G., Thums, A.: Safety Analysis of a Radio-Based Crossing Control System using Formal Methods. In: Schnieder, E., Becker, U. (eds.) 9th IFAC Symposium Control in Transportations Systems (2000)
6. Sato, K., Arai, H., Shimizu, T., Takada, M.: Obstruction detector using ultrasonic sensors for upgrading the safety of a level crossing. In: International Conference on Developments in Mass Transit Systems (1998) (Conf. Publ. No. 453)
7. Longrigg, P.: Railroad-Highway Vehicular Movement Warning Devices at Grade Crossings. IEEE Transactions on Industry Applications (March 1975)
8. Ku, B.-Y.: Grade-Crossing Safety. IEEE Vehicular Technology Magazine (September 2010)
9. Unmanned level crossings death trap for railways, Zee News, February 03 (2012)

MOFETA: A Network Architecture Based on MObile FEmtocells to Enhance Cellular Connectivity on TrAins

Patricia Noriega-Vivas, Celeste Campo,
Carlos Garcia-Rubio, and Alicia Rodriguez-Carrion

Department of Telematic Engineering
University Carlos III of Madrid, Spain
{pnoriega,celeste,cgr,arcarrio}@it.uc3m.es

Abstract. Nowadays the quality of the indoor coverage on trains is not enough to offer broadband services, mainly due to two factors: the high penetration losses caused by Faraday cage characteristics of the railcars and the high speeds reached. In this paper we aim to enhance the 3G indoor coverage on trains by setting up femtocells inside the railcars and establishing several wireless interfaces along the roof of the train. We design an architecture to connect the train to the fixed network in a transparent manner, so that the end users do not experience service disruptions during the journey. Finally, we deploy a testbed to evaluate the performance of the mobility management system based on test results.

Keywords: mobile femtocell, multihomed femtocell, femtocell, MOBIKE.

1 Introduction

Mobile computing is fast becoming a part of everyday life in which users demand being reachable anywhere and anytime. We spend much time traveling from one place to another, often by train. An important part of passengers is moving from air transport to high-speed trains due to the possibility of being able to get the most of their journey time, using their mobile devices connected to the Internet. Providing indoor coverage on trains directly with outdoor base stations may not be a good solution due to the high penetration losses caused by the Faraday cage characteristics of the railcars. This fact leads to a poor signal quality inside the train and offering broadband services is not always possible. Broadband access on trains could be achieved by installing base stations close to the railways, however this solution is not attractive for train operators due to the high investment needed to deploy such base stations. This issue has focused the research community effort on offering solutions that take advantage of the existing wireless infrastructure, in order to propose efficient methods to manage the mobility in a seamless way for the end user.

A. Vinel et al. (Eds.): Nets4Cars/Nets4Trains 2012, LNCS 7266, pp. 174–185, 2012.

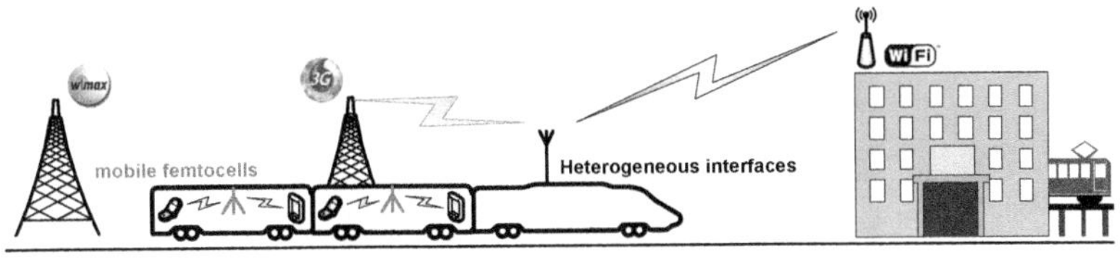

Fig. 1. Internet connectivity on trains

In recent years solutions have emerged to support Internet connectivity on trains relying on a single wireless technology. Aguado *et al.* [1] propose an architecture based on WiMAX since it is able to support mobility at speeds up to 500 km/h. Other approaches are based on satellite, IEEE 802.11 or emerging standards such as the IEEE 802.20. There are also approaches relying on several technologies instead of one. Rodríguez *et al.* [12] argue that a more efficient approach is to use a multitude of wireless technologies simultaneously and present a system combining several wireless interfaces. Most related works deploy Wi-Fi APs inside the train to connect to the Wi-Fi devices of the end users. However, users frequently have 3G devices such as mobile phones, tablets or UMTS dongles, which can also connect to the Internet and provide cellular connectivity.

Moreover, today's users are expecting seamless and ubiquitous access to voice and data services regardless of their location. In order to prevent users from being aware of mobility, it would be desirable offering the same coverage indoors and outdoors. In this paper we aim to provide on board 3G connectivity on trains by means of femtocells [4] (they will be presented in Section 2). Figure 1 shows the main idea behind this work. In the roof of the train several wireless interfaces are installed which provide connectivity with the outside world. Then, the signal is forwarded to the femtocells deployed inside the railcars which provide 3G connectivity to end users. This solution is responsible for selecting the most suitable technology at a given time according to several requirements in order to achieve the best performance. For instance, if the train is stopped in the train station it could connect to a Wi-Fi network but if it is on the move it could connect to cellular technologies such as WiMAX, UMTS, LTE or even satellite.

To that end, we have designed the MOFETA architecture (described in Section 3) to handle the mobility on trains. However, setting up femtocells on trains bring new challenges due to their fixed architecture. Mobility concerns for femtocell networks will be reviewed in Section 4. Section 5 presents a testbed that simulates some elements and functionalities presented by the MOFETA architecture. Finally, the results of this testbed will be discussed in Section 6.

2 Architecture for 3G Femtocells

Femtocells are low cost and low power cellular base stations typically used in homes and enterprises to improve indoor coverage and bandwidth. They are connected to the operator's core network by means of the user's broadband connection (DSL, cable...), hence they are envisaged to be deployed by users at

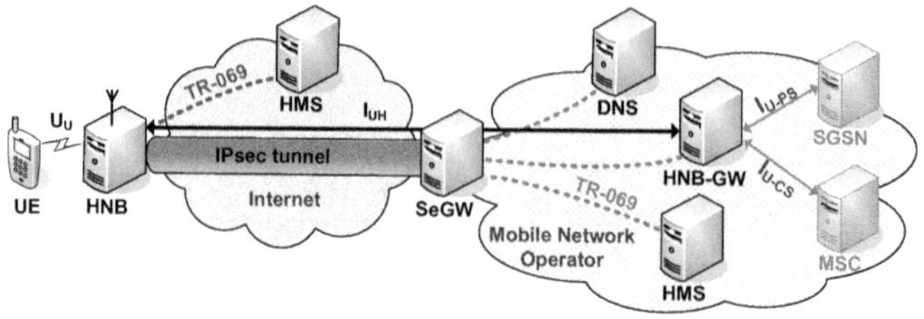

Fig. 2. 3G femtocell architecture overview

their premises. The new deployment model defined for femtocells requires new elements (see Figure 2). In this section, we briefly describe the new network elements and interfaces introduced by 3GPP in [17] for 3G femtocell networks.

- The **Home Node B (HNB)** is the femtocell. It serves User Equipment (UE) traffic by means of the UMTS U_U [16] interface and sends it to the core network over the new I_{UH} interface. It contains part, or all the functionality normally associated to a Radio Network Controller (RNC).
- The **Home Node B Gateway (HNB-GW)** terminates the I_{UH} interface and acts as a concentrator to aggregate a large number of HNBs. It is seen as a RNC by the core network and communicates with it using the existing I_{U-CS} and I_{U-PS} interfaces. [16]
- The **I_{UH} interface** connects the HNB with the HNB-GW and defines two new protocols in the control plane to address the differences between femtocell networks and the original I_U interface: The Home Node B Application Part (HNBAP) [19] and the RANAP User Adaptation (RUA) [18] protocols.
- The **HNB Management System (HMS)** is a management server that configures the HNB according to the operator's policy. It is composed of a TR-069 manager [15] and a file server. It performs location verification and assigns local access information to the femtocells.
- The **Security Gateway (SeGW)** terminates the IPsec tunnel established with the HNB, provides mutual authentication, encryption, data integrity and access to the operator's Intranet.

3 The MOFETA Architecture

In mobile and ubiquitous networks it is desirable that users do not experience fluctuations in the service quality when they are moving, i.e., they should not be aware of mobility. Nowadays, mobile users demand for the same quality in theirs own network than in mobile ones (e.g., travel by train). In this sense, efficient mobility management systems are needed to prevent the user from detecting changes in the quality of the services when moving.

In order to enable cellular connectivity on railway environments we propose the MOFETA architecture. It is based on the idea of using femtocells (henceforth referred to as mobile femtocells) inside the railcars to provide connectivity to

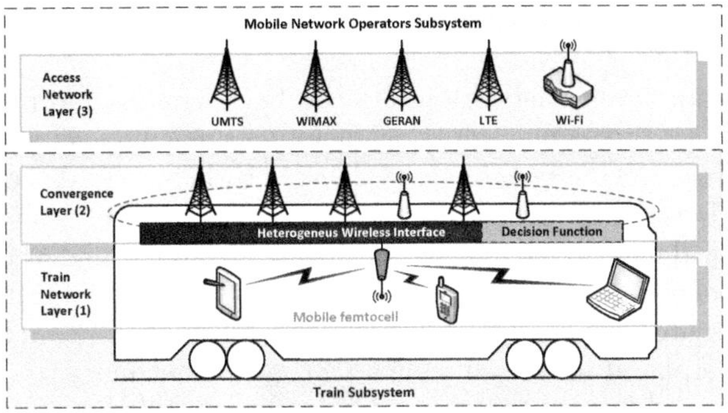

Fig. 3. MOFETA architecture overview

the passengers on board. We have designed this architecture to enhance the 3G connectivity inside the railcars and to support mobility between the train and the core network in a heterogeneous environment. This new architecture relies on the three differentiated layers described below (see Figure 3):

1. The **Train Network Layer (TNL)** consists of the set of mobile femtocells installed along the railcars and all the mobile users (passengers, train crew...) attached to them.
2. The **Convergence Layer (CL)** aggregates the traffic sent by femtocells in the TNL and forwards it to the Internet by means of the pool of wireless interfaces. Several wireless interfaces from several technologies are installed across the roof of the train to enable connectivity with the outside. This is meant as the Heterogeneous Wireless Interfaces (HWI) module, which acts as a gateway between mobile femtocells and the outdoors access points. Finally, the Decision Function (DF) is responsible to decide how and when a handover procedure must be triggered, i.e. it decides which technology is the most suitable at a given time to perform the handover.
3. The **Access Network Layer (ANL)** is composed by the outdoor wireless technologies available along the journey, e.g. UMTS, WiMAX, Wi-Fi...

This new layered architecture has potential advantages:

- It avoids multiple handovers. A single handover is required between the HWI and the ANL instead of perform a handover for each mobile node. Since only the HWI handles handover procedures, end users do not care about mobility.
- It combines several access technologies exploiting thus the environment diversity. Moreover, several interfaces from the same technology can be used to provide redundant links.
- It reduces the power required by mobile nodes due to the shorter distance between them and the femtocell.
- It makes a better use of the coverage area. Thanks to the antennas installed across the roof of the train, the signal quality received by them is better than inside the railcar, where high penetration losses occur [21]. This results in a longer use of the wireless link and therefore the number of handovers needed might be reduced.

4 Femtocells and Mobility

Femtocells include several network interfaces, both wireless and wired. The wireless interface is used for communications between the femtocell and the end user, e.g. mobile phones, PDAs or tablets. WCDMA, UMTS, WiMAX or LTE are the most popular femtocell-based solutions. Moreover, it is worth mentioning that there exist hybrid femtocells, able to provide both cellular and Wi-Fi connectivity. On the contrary, the wired interface is used to connect the femtocell with the mobile operator's core network over a fixed broadband connection. Hence, they are always placed in a fixed venue, i.e they cannot move.

Femtocells are also expected to be deployed in mobile environments such as buses, trains or airplanes in order to offer better coverage and bandwidth to passengers. However, supporting mobility on femtocells poses new challenges due to their fixed architecture. An important issue to deal with when moving femtocells is the IPsec tunnel established between femtocells and SeGWs (see Figure 2). The MOFETA architecture employs different wireless technologies to connect the train to the Internet according to decision procedures. During the journey, trains will be under the range of different wireless technologies. Due to handover procedures, changes in the IP address of the mobile femtocell are expected. This fact leads to the IPsec tunnel break and thus in service outage.

We already proposed improvements to the protocols defined by 3GPP for femtocell networks in [11]. In this work we also propose and evaluate the use of the MOBIKE protocol [6] between the femtocell and the core network (SeGW) in order to maintain the IPsec tunnel during handovers and thus providing continuity of service to the traffic conveyed by it.

MOBIKE is an extension to the existing IKEv2 protocol [7] to support secure mobility in IPsec tunnels. MOBIKE updates the IP addresses tied with IKEv2 and IPsec security associations (SA) without need for disconnecting existing SAs and establishing new ones, therefore continuity of service can be achieved. MOBIKE makes this possible managing the Security Association Database (SAD) and the Security Policy Database (SPD) [9] by using an internal API.

Femtocells and SeGWs already implement the IKEv2 protocol to establish and maintain security associations between them [20]. In that sense, our proposal would suppose minor changes in the current femtocell architecture since only femtocells and SeGWs would need to be upgraded to support the MOBIKE extension and thus handle the mobility in the IPsec tunnel.

In order to understand the operation of the MOBIKE protocol, we show in Figure 4 an example of tunnel reestablishment between two entities, an initiator (the entity that initiates the connection) and a responder. The initiator could be a mobile node with several interfaces, e.g. the mobile femtocell proposed in this paper, whilst the responder must be a stable node, e.g. a SeGW. Let us assume that the mobile femtocell (mHNB) is under the range of two wireless technologies so that it has obtained two IP addresses from them, IP_1 and IP_3. On the contrary, the SeGW is a fixed and stable node reachable at IP_2.

Communication using IKE always begins with IKE_SA_INIT exchange, where cryptographic algorithms are negotiated. Then, in the IKE_AUTH exchange,

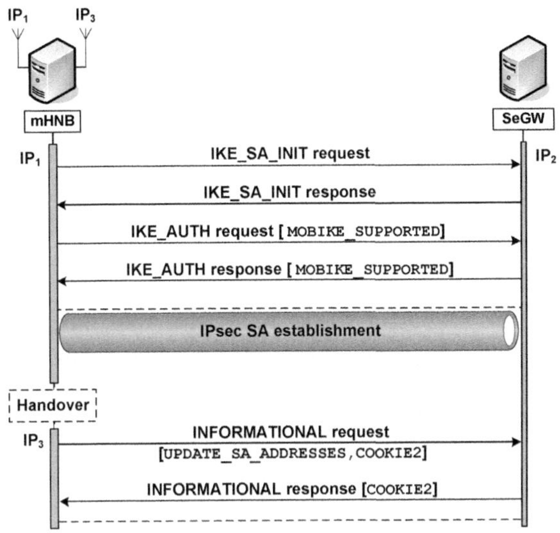

Fig. 4. MOBIKE handover procedure

peers transmit their identities, prove knowledge of the secrets and finally establish the first CHILD_SA for the ESP [8] protocol. MOBIKE introduces an additional payload to be used in the IKE_AUTH exchange. The `MOBIKE_SUPPORTED` payload is sent to inform the other peer that it supports the MOBIKE extension. Finally, if mutual authentication is successful, IKE and IPsec SAs are established between IP_1 and IP_2 and traffic exchanged between them is protected by ESP.

Now, let us assume that the initiator runs its decision function and decides to perform a handover to its secondary address, IP_3. Hereafter MOBIKE would proceed as follows: first, the initiator updates its SAs with its new address (IP_3) and then sends the `UPDATE_SA_ADDRESSES` payload from that IP to notify the responder. If the responder accepts the handover to that IP address, it updates its SAs and send back the same `COOKIE2` payload sent by initiator. This ensures the initiator that responder is replying to that concrete handover request.

5 Experimental Testbed

In this section, we aim to evaluate the feasibility of MOBIKE in order to use it in railway environments. To that end, we deploy a testbed to simulate the train scenario presented in Figure 1. We use three Linux hosts to emulate some elements in the femtocell architecture. The first host (initiator) emulates a mobile femtocell connected to three different wireless interfaces: WLAN, UMTS and GSM EDGE (GERAN). The second host (responder) emulates the SeGW, i.e. the entry point to the mobile operator's core network. To connect both elements we use a third host (router) which simulates network conditions. It simulates wireless technologies to connect to the femtocell and a fixed IP network to connect to

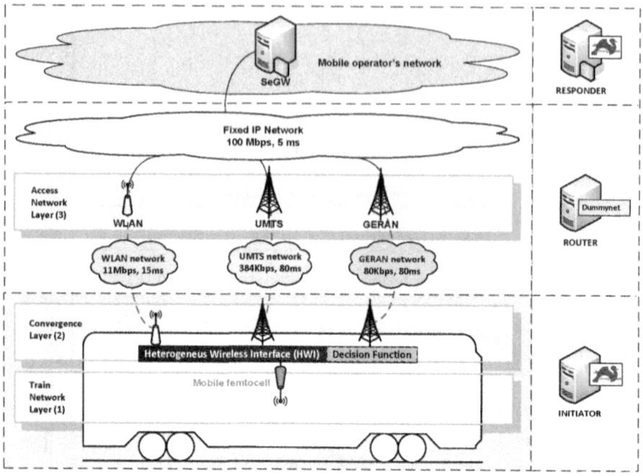

Fig. 5. Experiment overview

the SeGW. MOBIKE has been implemented on both initiator and responder to support mobility in the IPsec tunnel established between them. Figure 5 shows an overview of this testbed. The left column represents the functionality of each block whilst the right one shows the host that emulates such functionality.

We conduct experiments to evaluate the performance of MOBIKE when handovers occur. To that end, we measure the data traffic delay and losses during handovers to discuss later the validity of the solution depending on the concrete application. We define MOBIKE delay (D_M) as the time elapsed between the first data packet tunneled over the destination network and the last data packet tunneled over the home network when using MOBIKE to handle the mobility. Three different traffic sources are sent: CBR-12.2, CBR-45 and CBR-Max. CBR-12.2 has a send rate of 12.2 Kbps since it is the highest rate provided by the AMR codec [2], widely used in UMTS and GSM networks to encode voice traffic. Besides, CBR-45 sends 45 Kbps and CBR-Max sends with 70% of the wireless network bandwidth. Data is sent over UDP since it is the transport protocol defined for the I_{UH} interface in the user plane. Table 1 lists the properties of the three traffic sources employed in the experiment. CBR-45, CBR-Max traffic classes and the network parameters used (bandwidths and propagation delays for each network) are based on the experiment proposed by Ł Budzisz *et al.* in [3] in which authors study the feasibility of using the SCTP failover mechanism for handovers between WLAN, UMTS and GERAN networks.

There exist several IKEv2 open source implementations but only one includes the MOBIKE extension: StrongSWAN [13]. We install StrongSWAN in both femtocell and SeGW to handle the mobility between them. Besides, in order to simulate network conditions the router has installed Dummynet [5], a widely used tool that simulates bandwidth, delay and packet losses by configuring rules.

Table 1. Properties of the traffic types employed in the testbed

Traffic	Packet size	Send rate	Time between packets
CBR-12.2	32 Bytes	12.2 Kbps	20.983 ms
CBR-45	200 Bytes	45 Kbps	36.8 ms
CBR-Max-G	1410 Bytes	0.7×GERAN bandwidth (56 Kbps)	201.428 ms
CBR-Max-U	1410 Bytes	0.7×UMTS bandwidth (268.8 Kbps)	41.964 ms
CBR-Max-W	1410 Bytes	0.7×WLAN bandwidth (7.7 Mbps)	1.464 ms

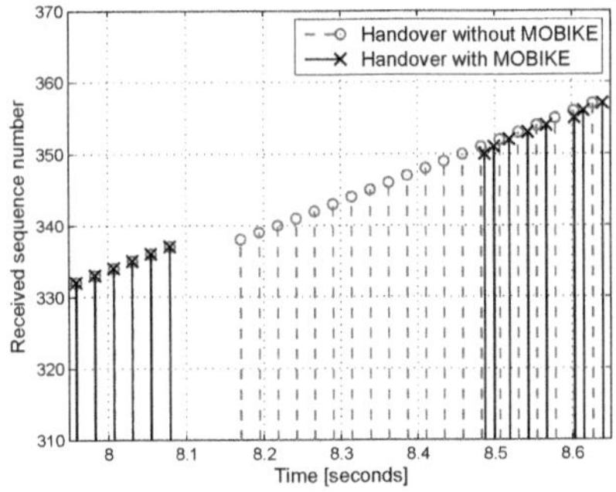

Fig. 6. WLAN to GERAN handover example (12.2 Kbps)

6 Result Analysis

To analyze the mobility support of MOBIKE, we create a test based on the one performed in [3]: first, data traffic is sent over the primary interface. Five seconds later, the secondary interface detects connectivity from another wireless technology and thus a secondary IP address is obtained. Three seconds later, the primary interface loses connectivity and as a result, a handover is triggered to the secondary interface. We repeat this test 30 times and then, we compute the mean and 95% confidence interval of the measured D_M and packet loss for each handover scenario. Specifically, we contemplate the following handover scenarios: UMTS to WLAN, UMTS to GERAN, WLAN to UMTS, WLAN to GERAN, GERAN to UMTS and GERAN to WLAN. This section presents the results according to three approaches, depending on the home network propagation delay (D_{hn}) and the destination network propagation delay (D_{dn}).

6.1 Approach 1: $D_{hn} < D_{dn}$

In this first approach, D_{hn} (15 ms) is lower than D_{dn} (80 ms). Hence, WLAN to UMTS and WLAN to GERAN handovers are grouped in this classification. Figure 6 illustrates a comparison between two handover examples when WLAN is the home network and GERAN the destination one. The red and dotted

Table 2. Approach 1: Results

Scenario	Traffic class	Average D_M [ms]	Average packet loss [packets]
WLAN	CBR-12.2	378.009±2.683	11.8±0.159
↓	CBR-45	392.014±0.005	7±0
UMTS	CBR-Max (7.7 Mbps)	380.743±0.732	68.866±0.305
WLAN	CBR-12.2	403.747±1.172	12±0
↓	CBR-45	437.864±5.417	7.33±0.179
GERAN	CBR-Max (7.7 Mbps)	526.564±7.463	72.833±1.503

bars represent the packets received by the responder ideally, i.e., there isn't any additional delay introduced by mobility management, but the delay added is produced by the inherent features of the networks crossed (transmission and propagation delays). The black bars represent the packets received by the responder when using MOBIKE to handle the mobility. In this case several delays are added, those produced by the networks crossed and the one produced by the MOBIKE implementation, StrongSWAN.

In that example, we send CBR traffic at 12.2 Kbps (e.g. a passenger voice call). At second 8 the handover is forced from the primary (WLAN) to the secondary interface (GERAN). In both cases, the last packet sent over the home network (#337) is seen at second 8.079. Ideally (red bars), packet 338 is received over GERAN 67.176 ms later than packet 337 due to the propagation and transmission delays introduced by GERAN. By contrast using MOBIKE, 12 packets are lost during the handover and hence the first packet received from the destination network (#350) arrives at second 8.482. Here, D_M takes 403.31 ms which is the time elapsed between packet 350 and packet 337 (black bars). Below, Table 2 shows the mean and 95% confidence interval of D_M and packet loss after repeat 30 times each test on each handover scenario and traffic class.

Wireless communication requirements for railway environments can by classified into 4 groups [10]. Class D communication is tied to passenger expectations so that the requirements depend on the user behavior. Train passengers would like to spend their journey time on surfing the Internet using their tablets, watching TV series on their laptops or talking on their cell phones. Depending on its nature applications have different requirements. Real time communications have strict requirements on delay, jitter and packet loss. For instance, voice and interactive video require delays not exceeding 150 ms whilst streaming-video applications are not delay sensitive due to the buffering (4 to 5 seconds are allowed) [14]. In that sense, the results obtained for WLAN to UMTS and WLAN to GERAN handoffs do not satisfy the delay constraints for voice and interactive video applications for any traffic class, preventing thus the use of the MOBIKE implementation (StrongSWAN) in those scenarios. On the contrary, all handover scenarios satisfy the streaming-video requirements for all traffic classes.

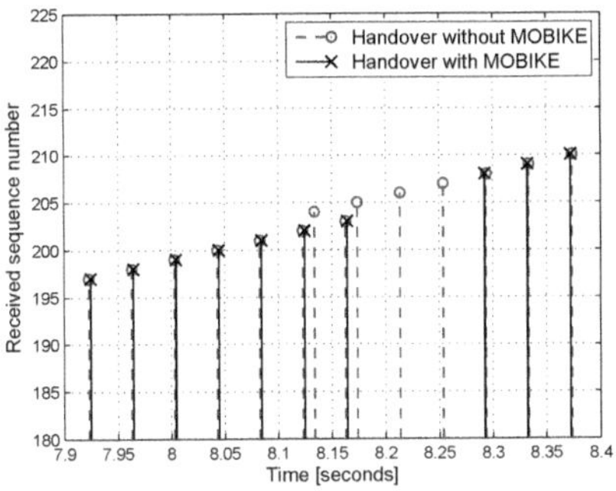

Fig. 7. UMTS to WLAN handover example (45 Kbps)

Table 3. Approach 2: Results

Scenario	Traffic class	Average D_M [ms]	Average packet loss [packets]
UMTS	CBR-12.2	100.543±0.852	6±0
↓	CBR-45	128.800±0.719	4±0
WLAN	CBR-Max (268.8 Kbps)	128.011±0.012	4±0
GERAN	CBR-12.2	94.1421.341	6±0
↓	CBR-45	104.937±0.641	4±0
WLAN	CBR-Max (56 Kbps)	208.582±1.503	1±0

6.2 Approach 2: $D_{hn} > D_{dn}$

The second approach represents those handovers in which D_{hn} (80 ms) is higher than D_{dn} (15 ms). Hence the first packets sent over the destination network could arrive earlier than the latest sent by the home network (i.e. they can anticipate). Figure 7 shows an UMTS to WLAN handover example when CBR-45 traffic is sent. In the ideal case, packet #203 arrives at second 8.162, 29.09 ms later than packet 204. Using MOBIKE we get a D_M of 128.50 ms, i.e. the time elapsed between packet 203 and 208, therefore 4 packets were lost due to handover.

In this approach most traffic classes provide average delays below 150 ms, therefore they satisfy the requirements for real time communications (voice, interactive video and streaming-video). Only in GERAN to WLAN handoffs sending CBR-Max traffic class we obtain an average delay above 150 ms. However that case satisfy the streaming-video requirements (see Table 3).

6.3 Approach 3: $D_{hn} = D_{dn}$

The last approach implies that both D_{hn} and D_{dn} have the same propagation delay (80 ms). Figure 8 represents a GERAN to UMTS handover example when CBR-Max traffic is sent (in this case, 56 Kbps). In the ideal case (red bars),

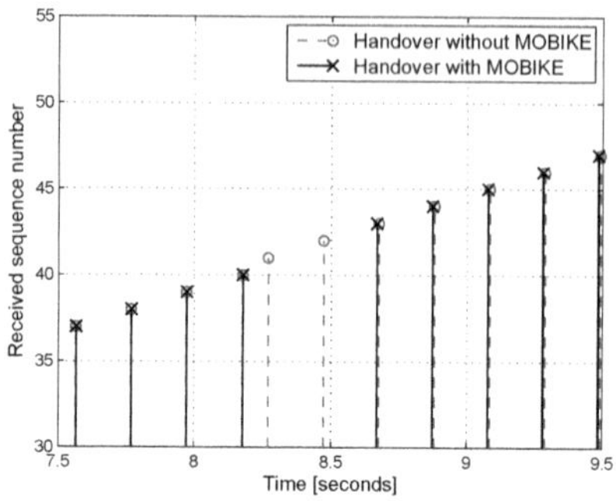

Fig. 8. GERAN to UMTS handover example (56 Kbps)

Table 4. Approach 3: Results

Scenario	Traffic class	Average D_M [ms]	Average packet loss [packets]
UMTS	CBR-12.2	336.555±1.399	12±0
↓	CBR-45	355.846±4.172	7.066±0.094
GERAN	CBR-Max (268.8 Kbps)	482.635±0.718	7±0
GERAN	CBR-12.2	286.782±2.894	11.166±0.141
↓	CBR-45	296.664±0.567	7±0
UMTS	CBR-Max (56 Kbps)	504.028±0.008	2±0

the first packet received over the destination network (packet 41) arrives earlier than if it had been sent over the home network, which has the same propagation delay. This is produced by the UMTS bandwidth, since it is greater than GERAN and therefore transmission delays are lower. Since the packet size used in this traffic class is large (1410 Bytes), transmission delays are not negligible. Using MOBIKE we compute 2 lost packets and a D_M equal to 504.02 ms (time between packet 40 and 43). Table 4 shows the mean values for D_M and packet loss for 30 tests for each handover scenario and traffic class. These results reveal that both handover scenarios are not feasible to voice and interactive video but they are for streaming-video applications.

7 Conclusion and Future Work

This paper presents a novel architecture to provide Internet access on trains by means of femtocells. Our solution considers several layers in order to separate the functionality. MOFETA assumes a pool of wireless interfaces on the roof of the train and femtocells inside the railcars to serve cellular users. This way of proceeding ensures better signal quality inside the train, since the signals are forwarded from the outside by means of cable, avoiding thus penetration losses.

MOBIKE has been proposed to handle the mobility in this architecture. We have evaluated some handover scenarios by using StrongSWAN, the only open source implementation of MOBIKE at this moment. We have presented the results for each scenario in terms of delay and packet loss, to argue the feasibility of StrongSWAN under concrete situations and applications.

In the future, we want to enrich the decision procedures by employing additional information. We anticipate to use level 2 information to know the signal quality of a given interface in order to aid in the decision-making procedures. For instance, link losses can be monitored to decide when a handover is necessary. Also we can take advantage of the fact that train routes are known in advance, to reserve resources in the next wireless network before initiate the handover.

Acknowledgment. This work has been supported by the Spanish Ministry of Science and Innovation, CONSEQUENCE project (TEC2010-20572-C02-01).

References

1. Aguado, M., et al.: Wimax on rails. IEEE Vehicular Technology Magazine (September 2008)
2. Digital cellular telecommunications system (phase 2+) (gsm); adaptive multi-rate (amr) speech transcoding (gsm 06.90 version 7.2.1 release 1998) (1998)
3. Budzisz, L., et al.: Towards transport-layer mobility: Evolution of sctp multihoming. Comput. Commun. 31, 980–998 (2008)
4. Chandrasekhar, V., et al.: Femtocell networks: a survey. IEEE Communications Magazine 46(9), 59–67 (2008)
5. Dummynet (January 2012), `http://info.iet.unipi.it/~luigi/dummynet`
6. Eronen, P.: Ikev2 mobility and multihoming protocol (mobike,rfc4555) (June 2006)
7. Kaufman, C., et al.: Internet key exchange protocol version 2 (ikev2,rfc5996)
8. Kent, S.: Ip encapsulating security payload (esp) (rfc 4303) (December 2005)
9. Kent, S.: Security architecture for the internet protocol (rfc 4301) (December 2005)
10. Kucharzyk, U.: Requirements for Wireless Technology on Rolling Stock. In: Strang, T., Festag, A., Vinel, A., Mehmood, R., Rico Garcia, C., Röckl, M. (eds.) Nets4Trains/Nets4Cars 2011. LNCS, vol. 6596, pp. 1–10. Springer, Heidelberg (2011)
11. Noriega-Vivas, P., Campo, C., Garcia-Rubio, C., Garcia-Lozano, E.: Supporting l3 femtocell mobility using the mobike protocol. In: Access 2011 (June 2011)
12. Rodriguez, P.: Mar: a commuter router infrastructure for the mobile internet. In: 2nd International Conference on Mobile Systems, Applications, and Services (2004)
13. Strongswan, `http://www.strongswan.org` (last accessed: December 9, 2011)
14. Szigeti, T., Hattingh, C.: End-to-End QoS Network Design: Quality of Service in LANs, WANs, and VPNs (Networking Technology). Cisco Press (2004)
15. Tr-069: Cpe wan management protocol v1.1, amendment 2 (December 2007)
16. Ts 25.401: Utran overall description (release 10) (June 2011)
17. Ts 25.467: Utran architecture for 3g home node b (hnb); stage 2 (rel 9) (June 2010)
18. Ts 25.468: Utran iuh interface ranap user adaption (rua) (rel 9) (September 2010)
19. Ts 25.469: Utran iuh interface hnb application part (hnbap) (rel 9) (September 2010)
20. Ts 33.320: Security of hnb/henb (release 10) (September 2010)
21. Verstrepen, W., et al.: Making a well-founded choice of the wireless technology for train-to-wayside data services. In: CTTE Conference (June 2010)

Author Index

GPSR Compliance

The European Union's (EU) General Product Safety Regulation (GPSR) is a set of rules that requires consumer products to be safe and our obligations to ensure this.

If you have any concerns about our products, you can contact us on ProductSafety@springernature.com

In case Publisher is established outside the EU, the EU authorized representative is:

Springer Nature Customer Service Center GmbH
Europaplatz 3
69115 Heidelberg, Germany

Batch number: 10844619

Printed by Printforce, the Netherlands